ST 31 – 180

SPECIAL
FORCES

HANDBOOK

JANUARY   1965

ST 31 - 180

SPECIAL TEXT 31 - 180

SPECIAL FORCES HANDBOOK

United States Army
John F. Kennedy
Special Warfare Center

January 1965 Edition

SPECIAL OPERATIONS PRESS
2013

ST 31 - 180
SPECIAL FORCES HANDBOOK
1965

This Edition Copyright © 2013 by Special Operations Press

Cover Art by: Special Operations Press

ISBN-13: 978-1481831383
ISBN-10: 1481831380

Proudly Printed in the
United States of America

Table of Contents

CHAPTER 4. AIR OPERATIONS

CHAPTER 5. WEAPONS

CHAPTER 9. MISCELLANEOUS

Tables

CHAPTER 1
GENERAL

I.   MISSION OF SPECIAL FORCES:

To plan and conduct unconventional warfare operations in areas not under friendly control.

To organize, equip, train, and direct indigenous forces in the conduct of guerrilla warfare.

To train, advise, and assist indigenous forces in the conduct of counterinsurgency and counterguerrilla operations in support of U.S. combat war objectives.

To perform such other special forces missions as may be directed or as may be inherent in or essential to the primary mission of guerrilla warfare.

II.   UNCONVENTIONAL WARFARE:

Unconventional Warfare is composed of the interrelated fields of:-

Guerrilla warfare.
Evasion and escape.
Subversion against hostile states.

III.   MISSIONS OF GUERRILLA FORCES:

Primary:
Interdict enemy lines of communication.
Interdict enemy installations and centers of war production.
and
Conduct other offensive operations in support of conventional military operations.

Supporting Tasks:
Intelligence.
Psychological warfare.
Evasion and escape.
Subversion against hostile states.

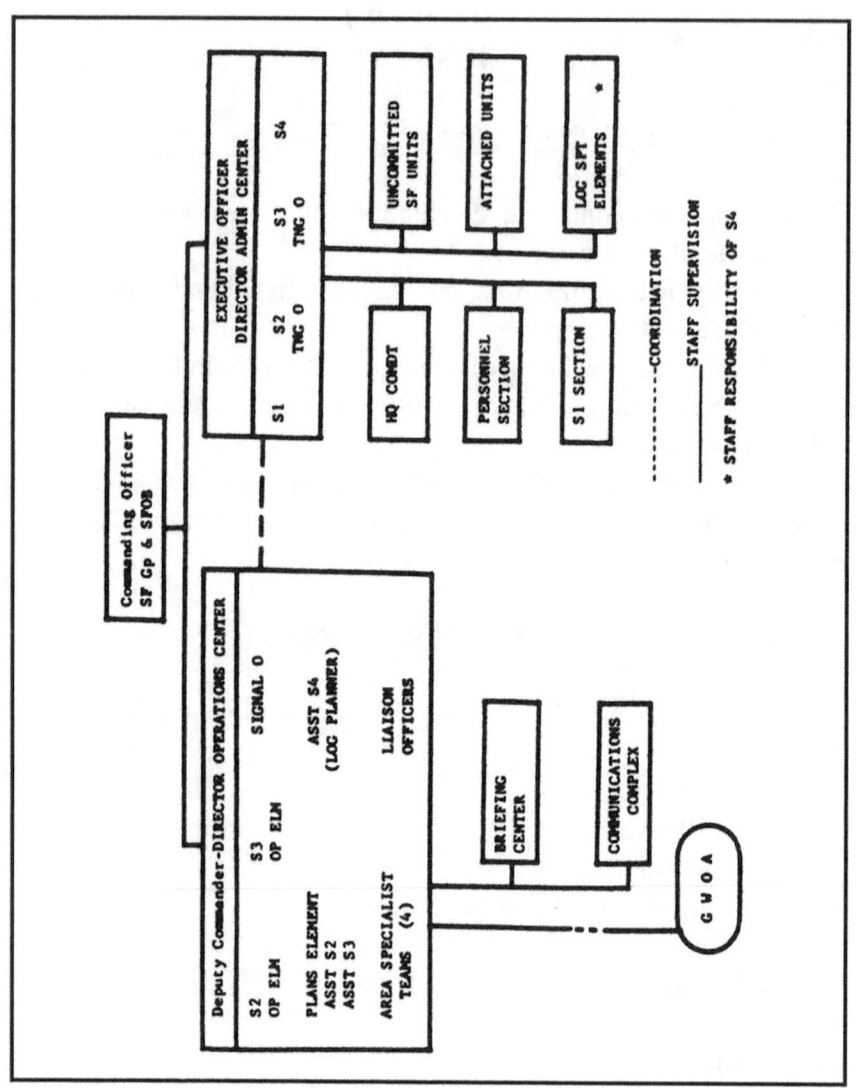

Figure 1 - Organization of SFG and SFOB

IV. COMPOSITION OF OPERATIONAL DETACHMENTS:

Operational Detachment A:

POSITION	RANK/GRADE
CO	Captain
XO	Lt.
OP SGT	E-8

INTEL SGT	E-7
LT WPNS LDR	E-7
HV WPNS LDR	E-7
MED SP	E-7
ASST. MED SP	E-6
RAD OP SUPV	E-7
RAD OP	E-5
DML SGT	E-6 C
MBT DML SP	E-5

Operational Detachment B:

CO	Major
XO	Captain
SMAJ	E-9
S1	Captain
S2	Captain
S3	Captain
S4	Captain
PREV MED SP	E-7
ADM SUPV	E-6
INTEL SGT	E-8
OP SGT	E-8
SUPPLY SGT	E-7
ASST SUPPLY SGT	E-6
LT WPNS LDR	E-7
HV WPNS LDR	E-7
DML SGT	E-7
CMBT DML	E-5
MEDICAL SP	E-7
RAD OP SUPV	E-7
RAD OPR (4)	E-5

Operational Detachment C:

CO	Lt Col
XO	Major
SMAJ	E-9
S1	Captain
S2	Captain

S3	Captain
S4	Captain
ADM SUPV	E-6
INTEL SGT	E-8
OP SGT	E-8
SUPPLY SGT	E-8
ASST SUPPLY SGT	E-7
RAD OP SUPV	E-7
RAD OP (4)	E-5
RAD REPAIRMAN	E-4

CHAPTER 2
TACTICS

ESTIMATE OF THE SITUATION

1. MISSION:

Mission assigned and analysis thereof to include sequence of task(s) to be performed and the purpose.

2. SITUATION AND COURSES OF ACTION:

Considerations affecting possible courses of action. Characteristics of the area of operation: weather, terrain, other. Relative combat power: enemy situation and friendly situation. Enemy capabilities. Own courses of action: Who, What, When, Where, Why, and How as appropriate.

3. ANALYSIS:

Select enemy capabilities.
List advantages and disadvantages. Course of action vs enemy capabilities. Course of action vs enemy capabilities.

4. COMPARISON:
Review and summary of advantages and disadvantages. Determination of significant advantages and disadvantages.

5. RECOMMENDATION/DECISION:

Formal statement of the course of action recommended/ adopted.

OPERATION ORDER

Task Organization: Includes the task subdivisions or tactical components comprising the command and reflects the unit of organization for combat.

1. SITUATION:

Enemy Forces: situation, capabilities. Indications.
Friendly Forces: missions and locations of higher adjacent, supporting and reinforcing units.
Attachments and Detachments: units attached to or detached from the unit issuing the order, for the operation concerned. Effective time of attachment or detachment is indicated when other than the time of the order.

2. MISSION:

Based on the order of the next higher headquarters and the commander's analysis of his mission, this paragraph contains a clear, concise statement of task(s) to be accomplished by the unit issuing the order and its purpose.

3. EXECUTION:

Concept of Operations.
Tactical mission of unit.
Coordinating Instruction: Tactical instructions and details of coordination applicable to two or more elements of the command.

4. ADMINISTRATION AND LOGISTICS:

Matters concerning supply, transportation, service, labor, medical evacuation and hospitalization, personnel, civil affairs and miscellaneous.

5. COMMAND AND SIGNAL:

Signal instructions and Information.
Command post and location of the commander.

ANNEXES:

Operation overlay.
Fire support plan.

DISTRIBUTION:

As Appropriate

II. PATROL LEADER'S ORDER

1. SITUATION:

Enemy forces: Weather, terrain, identification, location, activity, strength.
Friendly Forces: Mission of next higher unit, location and planned actions of units on right and left, fire support available for patrol, mission and routes of other patrols.
Attachments and Detachments.

2. MISSION:

What the patrol is going to accomplish?

3. EXECUTION:

(Subparagraph for each subordinate unit.)
Concept of operation.
Specific duties of elements, teams, and individuals.
Coordinating instructions: Time of departure and return. Formation and order of movement. Route and alternate route of return. Identification techniques used when departing and reentering the friendly area(s). Rallying points and action at rallying points. Location and actions at mission support sites. Actions on enemy contact. Actions at danger areas. Actions at objective. Rehearsals and inspections. Debriefing.

4. ADMINISTRATION AND LOGISTICS:

Rations.
Arms and ammunition.
Uniform and equipment. (State which member will carry and use.)
Method of handling wounded and prisoners.

5. COMMAND AND SIGNAL:

Signal. Signals to be used within the patrol. Communication with higher headquarters—radio call signs, primary and alternate frequencies, times to report and special code to be used. Challenge and password.
Command: Chain of command. Location of patrol leader and assistant patrol leader Information.

III. PATROL WARNING ORDER

The patrol warning order should consist of the following items of information.

A brief statement of the enemy and friendly situation.
Mission of the patrol.
General instructions. General and special organization. Uniform and equipment common to all, to include identification and camouflage measures. Weapons, ammunition, and equipment each member will carry. Who will accompany patrol leader on reconnaisance and who will supervise patrol members' preparation during patrol leader's absence. Instructions for obtaining rations, water, weapons, ammunition and equipment. The chain of command. A time schedule for the patrol's guidance. At a minimum, include meal times and the time, place, and uniform for receiving the patrol leader's order.

IV. TROOP LEADING PROCEDURE

Begin planning: Study terrain from map, sketch or aerial photo for: Critical terrain features. Observation and fields of fire. Cover and concealment. Obstacles. Avenues of approach. Make quick estimate of situation (thorough as time permits). Make preliminary plan.
Arrange for: Movement of unit (where, when, how). Reconnaissance (select route, schedule, persons to take along, use of subordinates). Issue of order (notify subordinate leaders of time and place). Coordination (adjacent and supporting units).

Make reconnaissance (examine the ground-see la above, if necessary changes preliminary plan).

Complete plan (receive recommendations, complete estimates, change preliminary plan as necessary, prepare order).

Issue order (include orientation on terrain if possible).

Supervise.

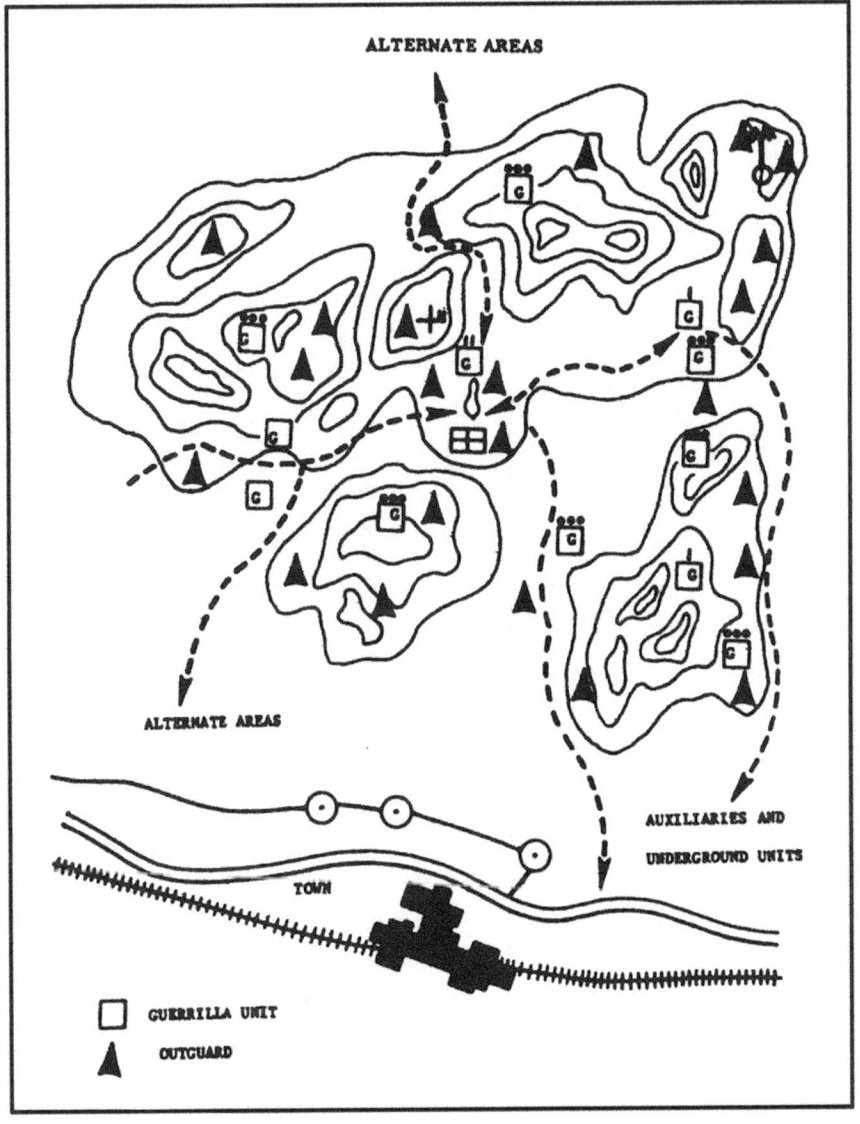

Figure 2 The Guerrilla Base and Alternative Areas System

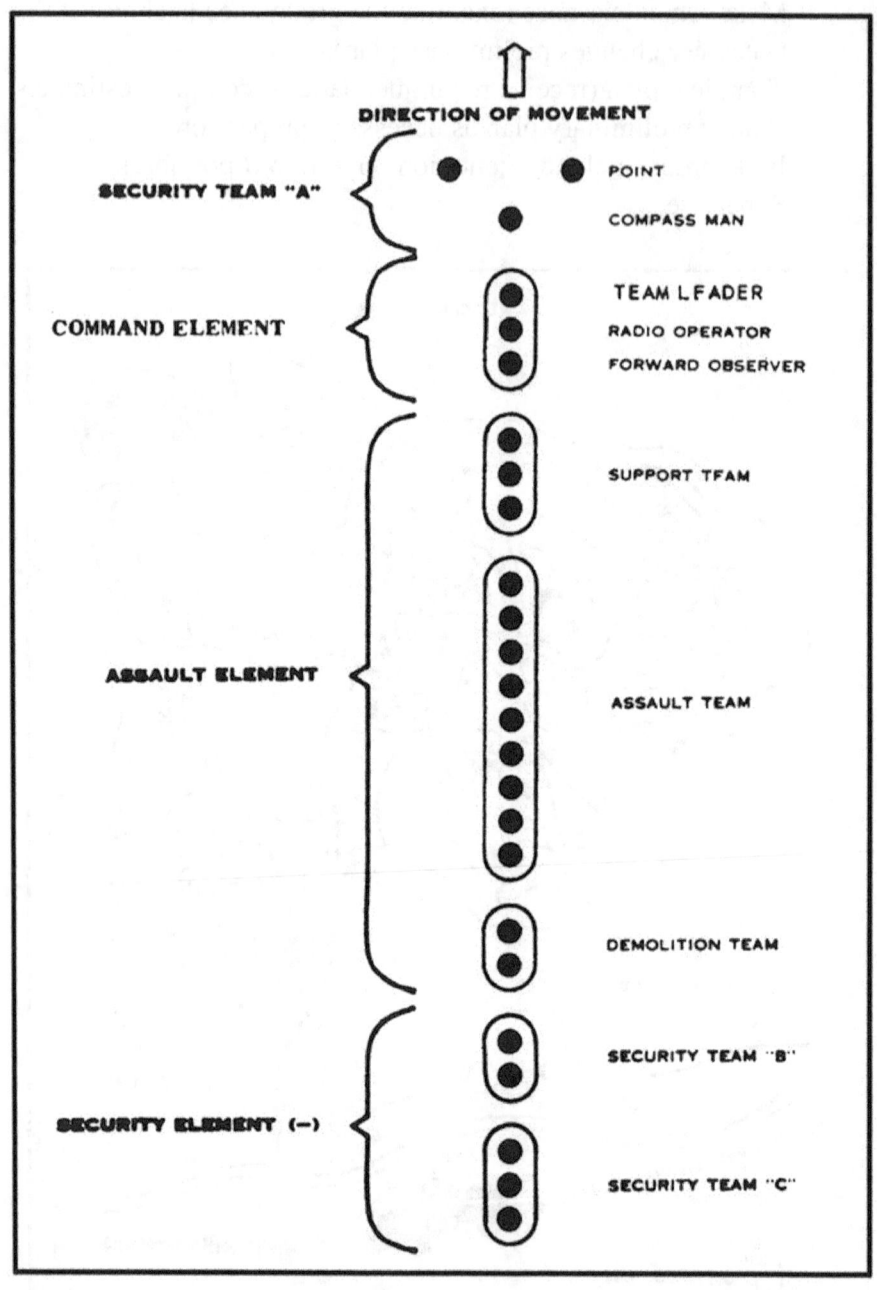

Figure 3. An example of the organization for movement of a raid force.

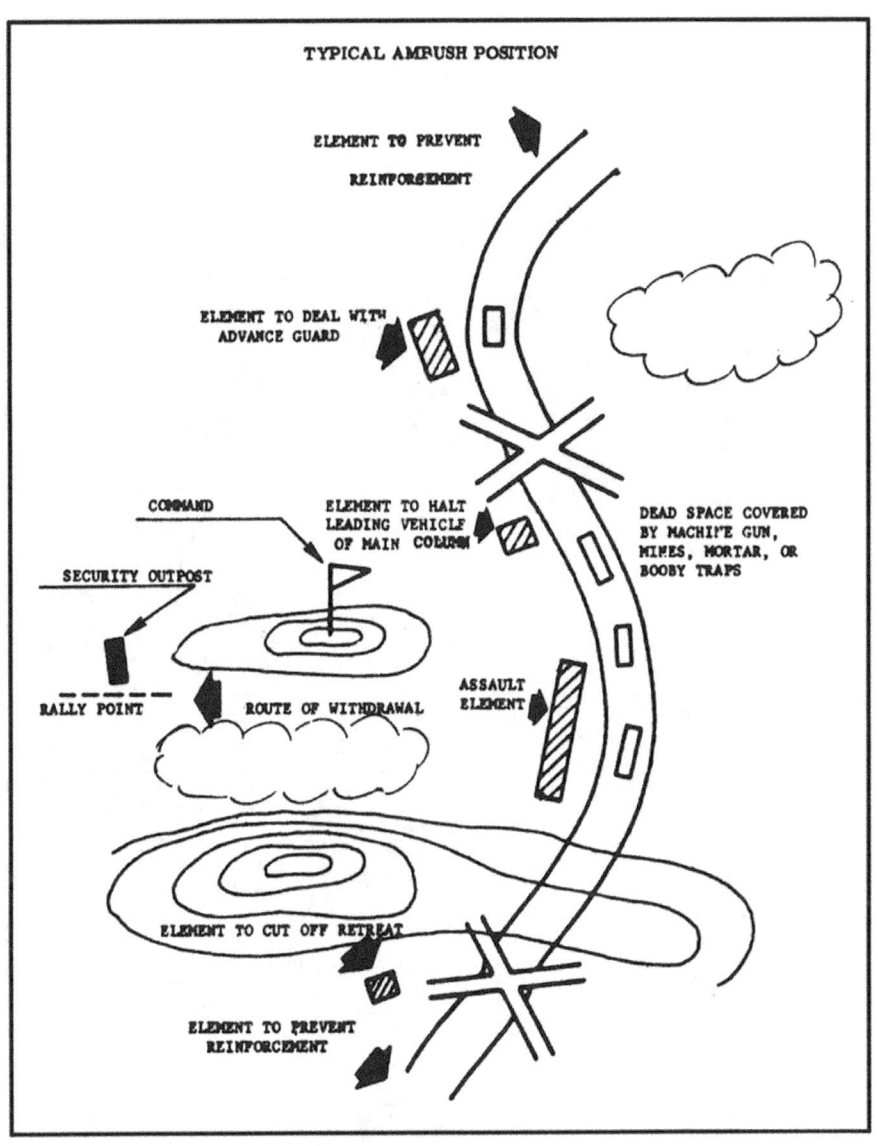

Figure 4. Typical Ambush Position

Figure 5. Conduct of Raid

20

CHAPTER 3
DEMOLITIONS

I.   INTRODUCTION

The following information pertaining to field engineering and demolitions is intended to supplement, but not to replace, that contained in FM 5-25, "Explosives and Demolitions," and FM 5-34 "Engineer Field Data." These field manuals, GTA 5-14, the Demolition Card and GTA 5-21, the Mine Card, are convenient references that should be obtained and used in conjunction with this section of the handbook.

	TNT	CYCLONITE	TETRYL	PETN	AMM NITRATE
USA	TNT	CYCLONITE *C-4	TETRYL *TETRYTOL	PETN *PENTOLITE *PRIMACORD	AMM NITRATE AMATOL
BRITISH	TNT *TROTYL	PLASTIC EXPLOSIVE OR *PE-2A	COMPOSITION EX OR C.E.	PETN *PENTOLITE *CORD TEX (DETCORD)	*AMMONAL *MONOBEL (AUST)
FRENCH	TOLITE				NITRATE D'AMMONIUM *TOLITE
GERMAN	PULL PULVER SPRENG MUN 02	*CYCLONITE *HEXOGENC 6 *PLASTIC *NIPOLIT		*KNALLZUND-SCHURR**	*AMMON SALPETER
ITALIAN	TRITOLO *TRITDO	*HENAGENE *T-4			
JAPANESE	CHA KATSUYAKY	KOSHITSUBKUYAKU *CYCLONITE *O-SHITSUYAKU	MELAYAKU	SHOE-J-YAKU	AMMON YAKU *SHONAYAKU *SHOAN *GOKUYAKU
RUSSIANw	TOL *TRYTYL	HEXOGEN *KAMNIKITE	TETP	TEN *DSH 1943**	*GROMOBOY *AMMONITE *DINOMANONK *MAISITE

*Compounded with other explosives   **Not known whether this is demolition explosive or a detonating cord.

Table 1. Principal Explosives of the World

STEEL, STRUCTURAL:

$$P = \frac{3A}{8} \text{ (in)}$$

$$K = \frac{1}{38} A \text{ (cm)}$$

STEEL, RODS, BARS, CABLES (2 inches or less hard carbon steel):

$$P = D^2 \text{ (in)}$$

$$K = \frac{D^2}{14} \text{ (cm)}$$

WOOD, EXTERNAL:

$$P = D^2 \text{ (in)}$$

$$K = \frac{D^2}{550} \text{ (cm)}$$

WOOD, INTERNAL:

$$P = D^2 \text{ (in)}$$

$$K = \frac{D^2}{3500} \text{ (cm)}$$

PRESSURE:

$$P = 2H^2 r \text{ (ft)}$$

Metric pressure formula
NOTE: When metric weights and measures are used, substitute breaching formula for the pressure formula.

BREACHING:

$P = R^3 KC$ (ft) (Add 10 percent if less than 50 pounds)

$K = 16 R^3 KC$ (m) (Add 10 percent if less than 22.5 KG)

C - Tamping factor for breaching (see demo card)

K - Material factor for breaching

MATERIAL	R IN FEET	R IN METERS	K FACTOR
ORDINARY EARTH	ALL VALUES	ALL VALUES	.05
Poor masonry shale, good timber & earth construction	All values	All values	.23

Table II - 1. Basic Demolition Formulas

MATERIAL	R IN FEET	R IN METERS	K FACTOR
Good masonry, ordinary concrete, rock	Less than 3	Less than 1	.35
	3 to 5	1 to 1.5	.28
	5 to 7	1.5 to 2	.33
	More than 7	More than 2	.28
Reinforced concrete (however will not cut reinforcing steel)	Less than 3	Less than 1	.70
	3 to 5	1 to 1.5	.55
	5 to 7	1.5 to 2	.50
P = Amount of TNT, in pounds, required for an external charge. For relative effectiveness of other external charges, see Table IIA.			

Table II - 2. Basic Demolition Formulas (Cont...)

Name	Principal use	Smallest cap required for detonation	Velocity of detonation (ft per sec)	Relative effectiveness as external charge (TNT = 1.00)	Intensity of poisonous fumes	Water resistance
TNT	Main charge, booster charge, cutting and breaching charge, and general use in combat areas.		21,000	1.00	Dangerous	Excellent
Tetrytol			23,000	1.20	Dangerous	Excellent
Composition C3		Military blasting cap, electric or nonelectric.	26,000	1.34	Dangerous	Good
Composition C4			26,000	1.34	Slight	Excellent
Ammonium Nitrate (Military cratering charge).	Cratering and ditching (in container).		11,000	0.42	Dangerous	Good, of container is not ruptured)
Military Dynamite M1	Land clearing, quarry and rock cuts and general use		20,000	0.92	Dangerous	Good
Straight Dynamite 40% (Commercial) 50% 60%	Land clearing, quarry and rock cuts, and general use.	No. 6 commercial cap, electric or nonelectric.	15,000	0.65	N/A	Good of fired within 24 hrs
Ammonia Dynamite 40% (Commercial) 50% 60%	Land clearing, cratering, quarrying and general use in rear areas.	No. 6 commercial cap, electric or nonelectric.	9,000 11,000 12,000	0.41 0.46 0.53	Dangerous	Poor
Gelatin Dynamite 40% 50% 60%			8,000 9,000 16,000	0.42 0.47 0.76	Slight	Good
PETN	Detonating cord	No. 8 commercial cap*	21,000	N/A	Slight	Good
	Blasting caps	N/A	N/A	N/A	N/A	N/A
TETRYL	Booster charge	Military blasting cap*	23,400	1.25	Dangerous	Excellent
	Blasting caps	N/A	N/A	N/A	N/A	N/A
Composition B	Bangalore torpedo and shaped charges.	Military blasting cap, electric or nonelectric.	25,500	1.35	Dangerous	Excellent
Black Powder	Time fuse	N/A	N/A	N/A	Dangerous	Poor
RDX	M6 and M7 blasting caps.	N/A	N/A	N/A	N/A	N/A

*Electric or nonelectric

Table IIA. Characteristics of Explosives

II.   RAIL CUTS

While single rail cuts have a harrassing or nuisance value, we will usually be concerned with cuts designed to derail a train. In order to insure the derailment of a modern locomotive it is necessary to re-move a length of rail equal to the length of the fixed wheel base of the locomotive. The weight of a locomotive is counterbalanced in such a way that the removal of rail less than the length of the fixed wheelbase may not result in derailment.

a.   Twenty-foot gap Technique. World War II experience and related tests have established that a charge sufficient to remove 20 feet of rail will result in positive derailment of a locomotive under most operational situations. The most effective cut is on the outside rail of a curve. Where two or more tracks parallel, derailment should be made in such a manner that a train, when wrecked rounding a curve on the inside track, will obstruct all tracks. When derailment is attempted on a straight stretch of a multiple track line, attack should always be made on an inside rail. Note that in all cases only one of the two rails of a track is attacked.

b.   The Derailment Charge requires three quarter pounds of plastic explosive, either CS, C4 or their equivalent, to cut the standard rail (80 lbs. per yard). One-third of the standard issue plastic demolition block is a convenient unit of measure. A series of three, quarter-pound charges is arranged on the web of the rail as diagrammed in figure 1. The series of charges should not bridge a fishplate. One charge is placed directly over each tie on the selected 20 feet of rail. This will result in removing all rail, at least partially breaking the ties directly under the rail, and creating some minor cratering of the roadbed ballast. Standard center tie spacing is 22 1/2 inches; however, variations run from 18 inches on up to 3 feet. Lacking specific information on the tie spacing, the distance between prepared charges is based on 18 inch measurement which results in placing 15 of the three quarter pound charges for each derailment series on a continuous detonating cord main line. The detonating cord main line to which the individual charges are attached is 28 feet in length to provide approximately a foot tail at either end for quick attachment of a firing system. A triple roll knot for each three quarter pound charge is fixed on the main line as diagrammed in figure 2. These knots are arranged roughly on 18 inch center to coincide with the anticipated tie spacing. They are arranged to insure a snug continuous contact with the main line but loose enough to slide; thus making it possible to make on target adjustments for variations in tie spacing. The individual three quarter pound charges are firmly molded around each triple roll knot. They must be sufficiently wrapped to withstand the necessary rough handling in bringing them on target and to also insure that the charge and knot will slide as an integral unit.

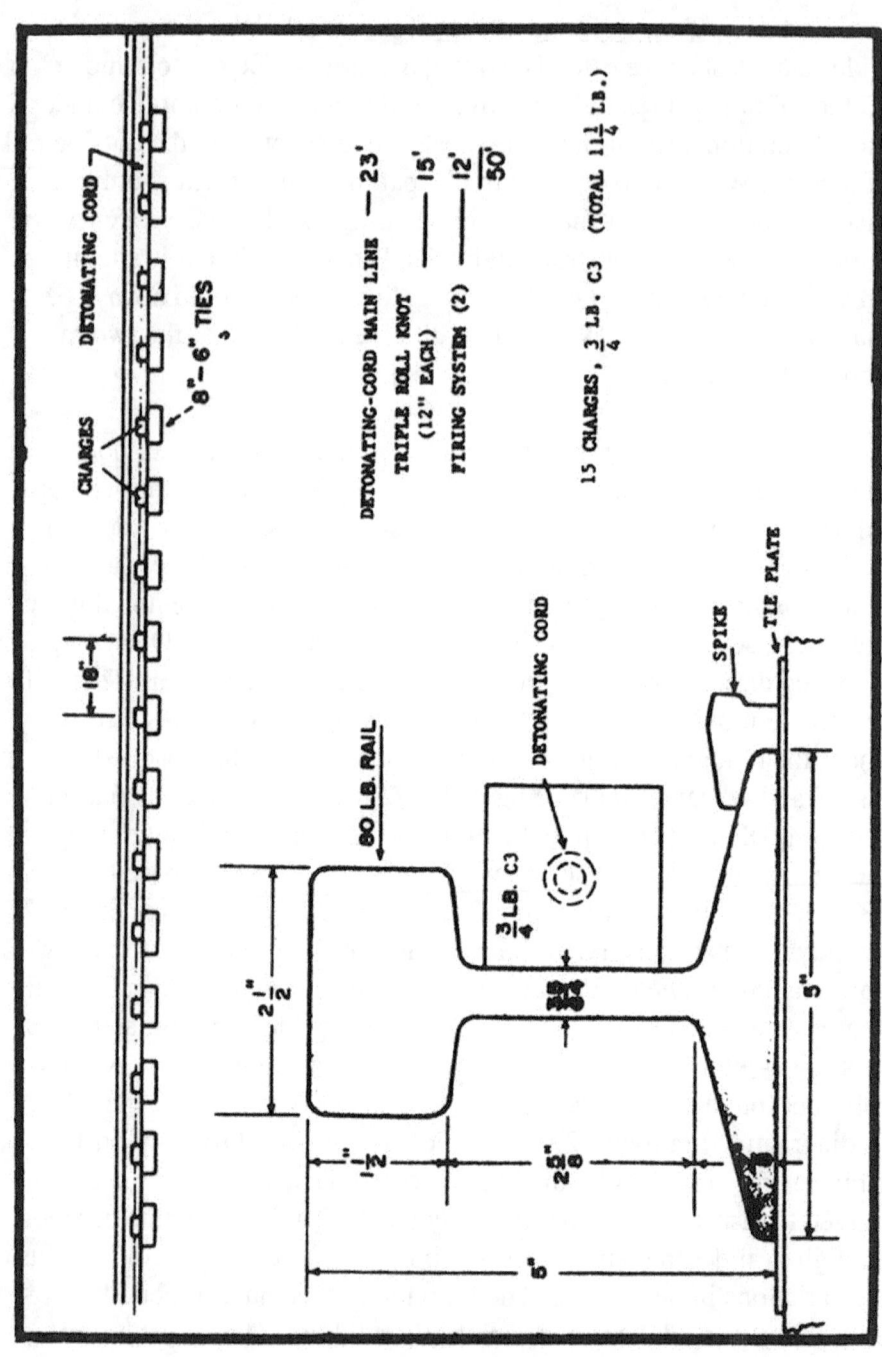

Figure 6. Hasty derailment charge.

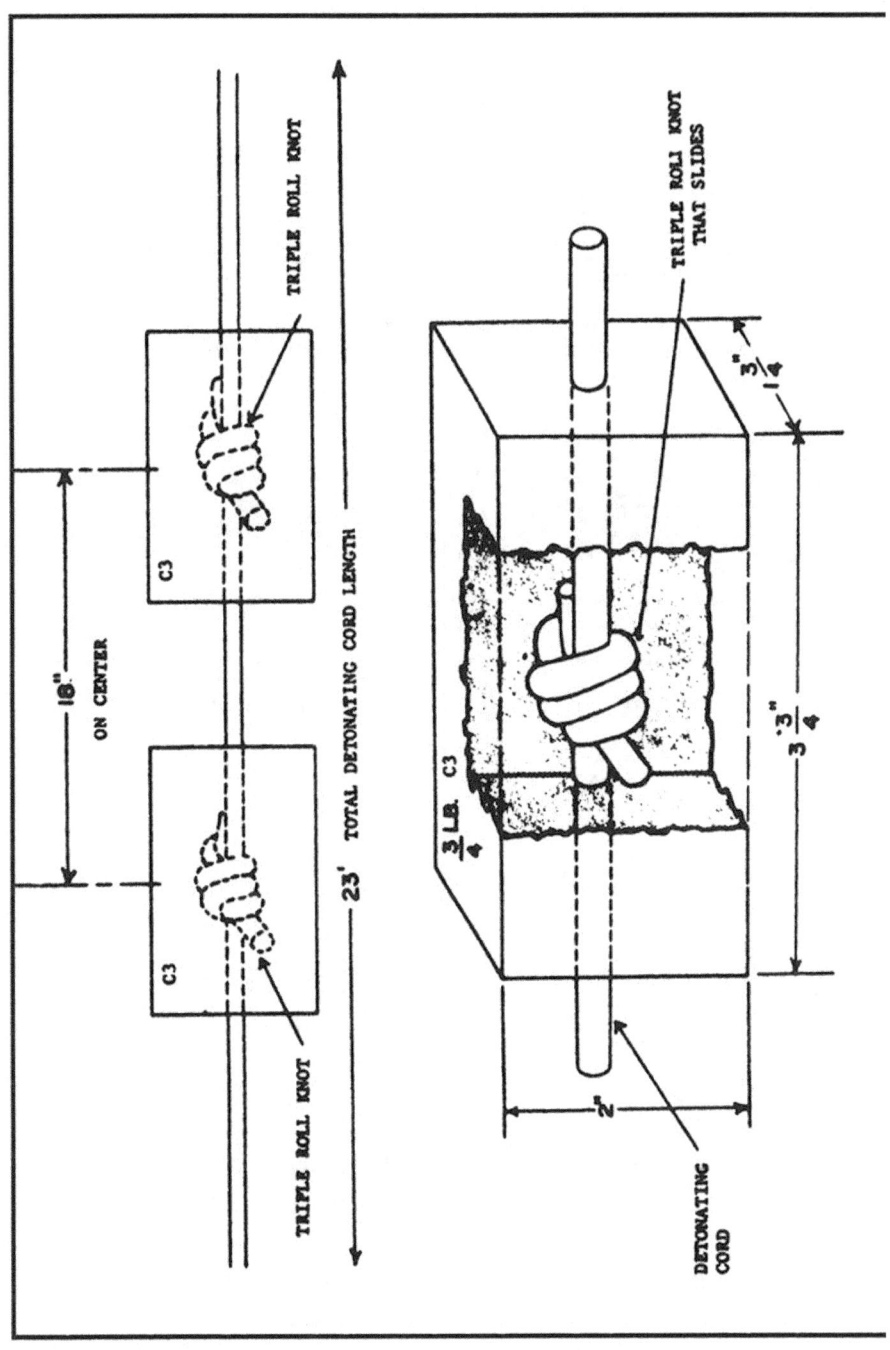

Figure 7. Hasty derailment charge showing use of detonating cord.

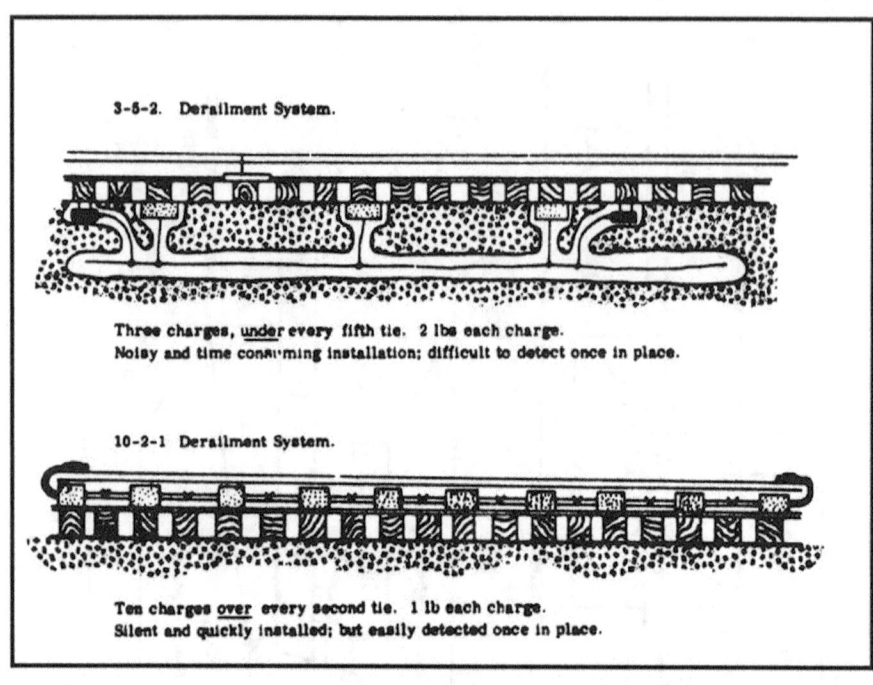

3-5-2. Derailment System.

Three charges, under every fifth tie. 2 lbs each charge.
Noisy and time consuming installation; difficult to detect once in place.

10-2-1 Derailment System.

Ten charges over every second tie. 1 lb each charge.
Silent and quickly installed; but easily detected once in place.

Figure 8. 3-5-2 & 10-2-1 Derailment Systems.

c.   <u>Firing Systems</u>.

A standard electric firing system is best for continuous and immediate control over initiating the charge. A standard nonelectric system may also be used and timed to insure that the charge explodes just in front of the train; however, both these systems require the presence of an agent at the scene of operations.

All the military booby-trap firing devices can be used to initiate the charge through the movement of the oncoming train. Home made firing devices employing the mechanical principles of the military issue booby-traps can be employed. An electrical blasting cap system may be activated with a flashlight battery used as a simple, fieldimprovised switch that is closed by the movement of the train. In all cases the firing system is set up to initiate the charge immediately in front of the oncoming locomotive, not under the locomotive. Eighty pound or less rail (5 inches or less in height) takes 1/2 pound to cut. Over 80 pound rail (over 5 inches in height) takes 1 pound to cut.

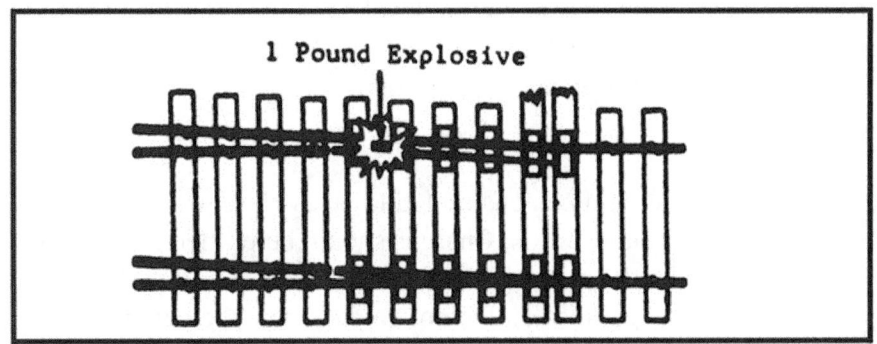

Figure 9.

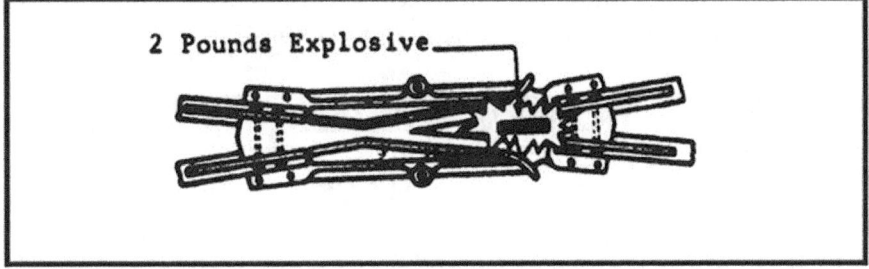

Figure 10.

Figures 9 &10. Junction Destruction.

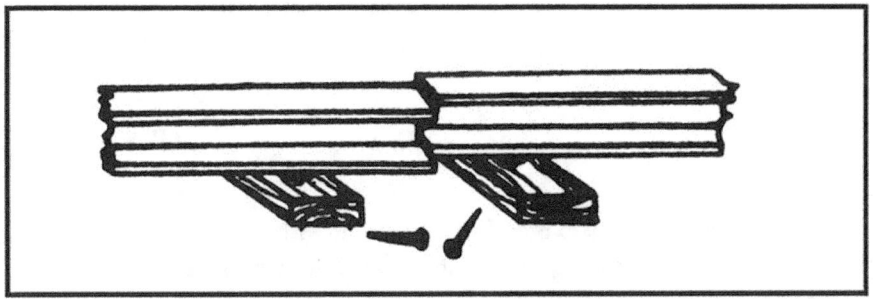

Figure 11. Manual Track Distortion

(With a jumper wire, provide a path for the electrical current passing through most rails. The wire that is normally between rails will be broken by this manual displacement.)

    d.   Only plastic explosive should be used, either C3 or C4. Information has been developed for breaching reinforced concrete targets from 1 through 8 feet in thickness. For maximum effect, the charge should be placed a distance equal to the thickness of the target above the base (or above the ground level). Charges placed at the base of a slab will still work but in study they produced craters 23 percent smaller than those placed above the ground.   Figure 1. Hasty derailment charge.

A charge should be constructed to be as close to square as possible to yield optimum results. Charges should be primed either from one corner or from the exact rear center. Close contact with the target is required for the best results. Do not deviate from the charge thickness indicated below. Use the M-37 kits as issued when possible to facilitate securing the charge in place. If it is necessary to cut the block, cut them with care so that the density of the explosive is not affected.

CONCRETE (IN FEET)	THICKNESS (METERS)	CHARGE SIZE (USE C4 BLOCKS)	CHARGE THICKNESS
1	.3	2	1 Block (2")
2	.61	4	1 Block
3	.91	7	1 Block
4	1.22	20	1 Block
		(USE M-37 KITS)	
5		6	2 Blocks (4")
6		8	2 Blocks
7		12	2 Blocks
8		20	2 Blocks

NOTE: Using the standard breaching technique with an untamped charge above the ground would require 124 lbs of TNT to breach a 4 foot wall. Using the above technique, it would require 50 lbs of C4. For a 7 foot target the standard method uses 517 lbs of TNT. This method uses 240 lbs of C4.

Table III. Charge Sizes.

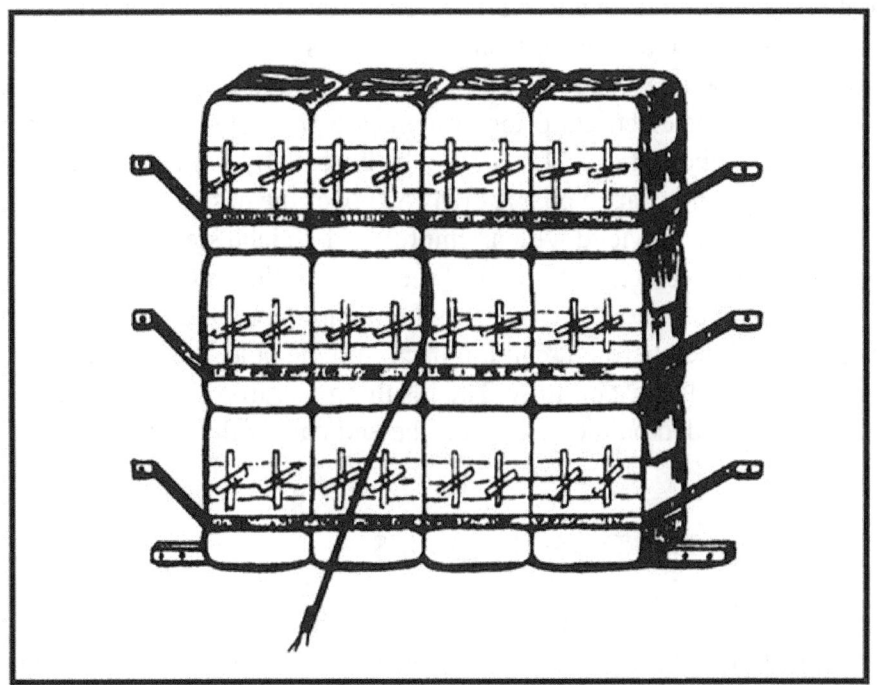

Figure 12. Square Charge

III.   CRATERING TECHNIQUE.

A delay cratering technique has been developed that produces excellent results, and it should be considered if time and materials are available. The charges themselves should be either the standard 40-pound cratering charge, or 30 to 40 pounds of C4 (depending to some extent on the depth and diameter of the bore holes). Depth of the holes should be 4 or 5 feet.

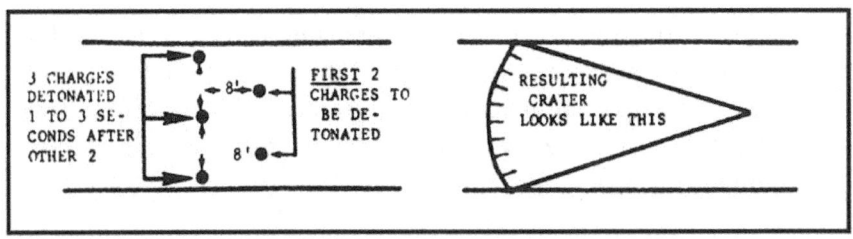

Figure 13. Cratering Techniques.

The line of two charges above should be primed to be deto-nated simultaneously. The line of three charges should all be primed to detonate simultaneously after the correct delay. The line of three charges should be detonated from one to two seconds after the first two charges detonate. The delay can be achieved in a variety of ways, but two separate electrical firing systems are probably the easiest. An expedient method of quickly sinking the bore holes themselves is to set up five 15 pound shaped charges (M2A3) over the desired locations. They should be provided with an improvised 30 inch standoff and all be detonated together. The effect of the delay in the cratering operation is to begin to move a large amount of earth around the first two charges, and then before it can fall back into the hole, the second line of charges displaces it entirely. The resulting teardrop shaped crater is very steep sided on the blunt end (the end having the three delay charges).

IV.   IMPROVISED DEVICES.

Bangalore torpedos, if available, can be extremely effective if employed in an antipersonnel role. Best results are obtained if the Bangalore is planted upright in the ground so that the fragmentation effect will radiate out in 360 degrees.

The fragmentation hand grenade is a versatile weapon that lends itself to a wide variety of booby trapping actions. One of the simplest booby traps is the grenade-in-a-can. The shipping container or can is affixed to a tree or other permanent object. The grenade, with pull ring removed is placed in the can so that the arming lever is held down by the can. A string or wire is then so placed that the victim will pull the grenade from the can, releasing the lever and detonating the grenade.

Improvising electrical booby trap firing devices. Each of the following simple booby traps can be used in conjunction with a wide variety of casualty producing charges, from the 3.5 inch rocket, fired by expedient electrical means, to the bangalore torpedo primed to be detonated in an antipersonnel role.

Open loop. The open loop arrangement shown is the ONLY break in an otherwise complete electrical circuit. A wide variety of actions on the part of the victim could result in pulling the two bare ends of the wire together.

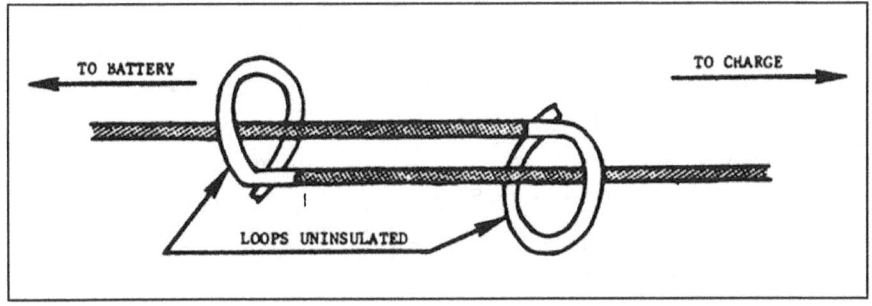

Figure 14. Open Loop.

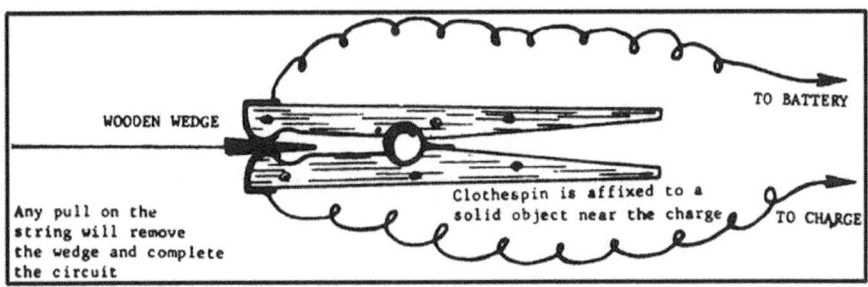

Figure 15. Clothespin.

Expedient firing of 3.5-inch Rocket. The following technique is one method for firing the rocket electrically. Either the cardboard shipping container or a V-shaped wooden trough may be used as an expedient launcher, with the trough being preferred if available. The rocket is prepared for electrical firing by locating the two wire in the nozzle and fin assembly that are coated with clear plastic. (The other green, red, and blue wires are disregarded). After scraping the ends of the clear plastic wires, to provide a good contact for splicing into the firing wire, the connection is made and preferably taped. Experience has indicated that the railing splice is the preferable splice to be used without any adverse effect on rocket accuracy.

The bare-riding safety band is removed and the rocket is placed on the trough so that the bore-riding safety will face a side of the trough during firing. The shorting clip is removed, the rocket is aimed, an electrical power source is provided and the rocket is fired.

Obviously results comparable to those obtained by using the launcher should not be expected. As with all expedient demolition work, trial and error experimentation is stressed. An experienced demolitionist can reliably hit a 55 gallon drum, a relatively small target, up to a range of 40 to 50 yards. In an antitank role, satisfactory results could be expected up to 150 yards. The rocket firing can be controlled by the operator, or can be effected by a wide variety of electrical booby trapping techniques. This expedient use of the rocket of course lends itself equally well to employment in an antipersonnel role.

Power sources can be a 10-cap blasting machine or any of the following dry cell batteries: BA-317/U, BA-270/U, BA-279/U or combinations of the BA-30/U.

Safety precautions should include all of those associated with electrical firing as outlined in FM 5-25, Explosives and Demolitions. Although it would be an extremely rare occurrence, we should operate on the assumption that the rocket may blow up on the launching site and take appropriate precautions to prevent injury from such an accident.

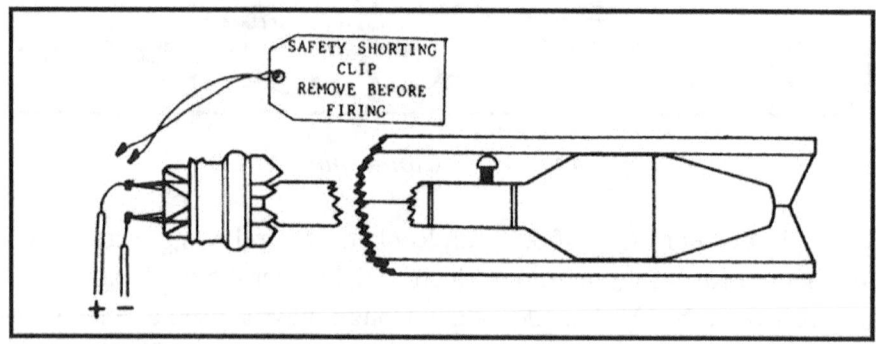

Figure 16. 3.5 Rocket in Firing Trough - electrical firing.

Expedient firing of 3.5 rocket nonelectrically.

Remove all wires from fin assembly.

Remove the plastic cone from fin assembly.

Place matchheads or other burning material in contact with the ends of the sucks of propellant.

Tape matches around end of fuze.

Place fuze flush against perforated disc, and among matchheads already in the nozzle.

Remove the bore-riding safety band and place the bore-riding safety pin in depressed position against sides of improvised firing platform.

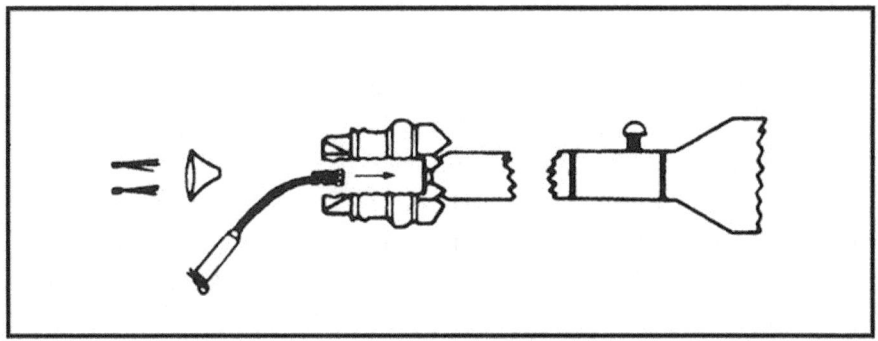

Figure 17. 3.5 Rocket fired nonelectrically.

V.   ADVANCED TECHNIQUES.

Charges constructed employing advanced techniques generally produce more positive results while using less explosive than required by conventional or standard formulas. Disadvantages of advanced technique charges are that they usually require more time to construct and once constructed they are usually more fragile than conventional charges. Following are rules of thumb for various charges and the targets they are designed to destroy.

Saddle Charge. This charge can be used to cut mild steel cylindrical targets up to 8 inches in diameter. Dimensions are as follows: The short base of the charge is equal to one-half the circumference. (Note that previously published dimensions called for three times the base, rather than twice the base.) Thickness of the charge is 1/3 block of C3 or C4 for targets up to 6 inches in diameter: use one-half block thickness for targets from 6 to 8 inches in diameter. Above 8 inches in diameter, or for alloy of steel shafts, use the diamond charge. Prime the charge from the apex of the triangle, and the target is cut at a point directly under the short base by cross-fracture. Neither the saddle nor diamond will produce reliable results against non-solid targets, such as gun barrels. These charges benefit from prepackaging or wrapping, providing that no more than one thickness of the wrapping material is between the charge and the target to be cut. Heavy wrapping paper or aluminum foil are excellent, and parachute cloth may be used if nothing else is available. (See figure 18.).

Diamond Charge. This charge can be used to cut hard or alloy steel cylindrical targets of any size that would conceivably be encountered. It has reliably been used, for instance, against a destroyer propeller shaft of 17 inch diameter. Dimensions are as follows: The long axis of the diamond charge should equal the circumference of the target, and the points should just touch on the far side. The short axis is equal to one-half the circumference. Thickness of the charge is 1/3 thickness of a block of C3 or C4. To prime the charge, both points of the short axis must be primed for simultaneous detonation. This can be accomplished electrically or by use of equal lengths of detonating cord, with a cap crimped on the end that is inserted into the charge. As detonation is initiated in each point of the diamond and moves toward the center, the detonating waves meet at the exact center of the charge, are deflected downward, and cut the shaft cleanly at that point.

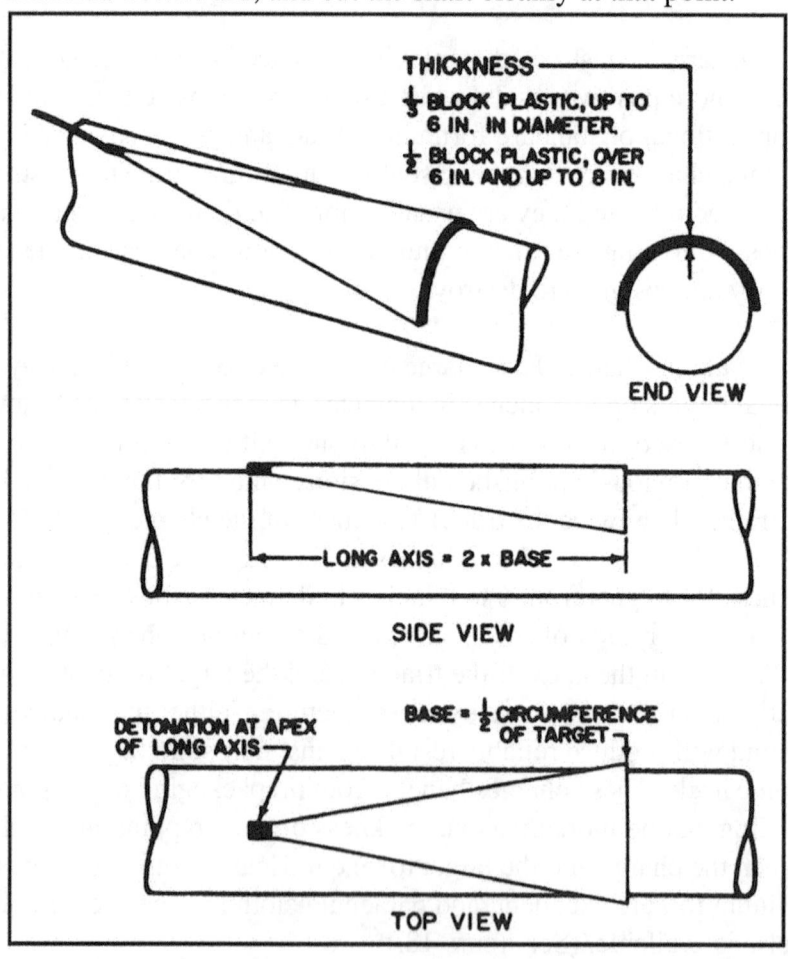

THICKNESS

$\frac{1}{3}$ BLOCK PLASTIC, UP TO 6 IN. IN DIAMETER.

$\frac{1}{2}$ BLOCK PLASTIC, OVER 6 IN. AND UP TO 8 IN.

END VIEW

LONG AXIS = 2 x BASE

SIDE VIEW

DETONATION AT APEX OF LONG AXIS

BASE = $\frac{1}{2}$ CIRCUMFERENCE OF TARGET

TOP VIEW

Figure 18. Saddle Charge.

36

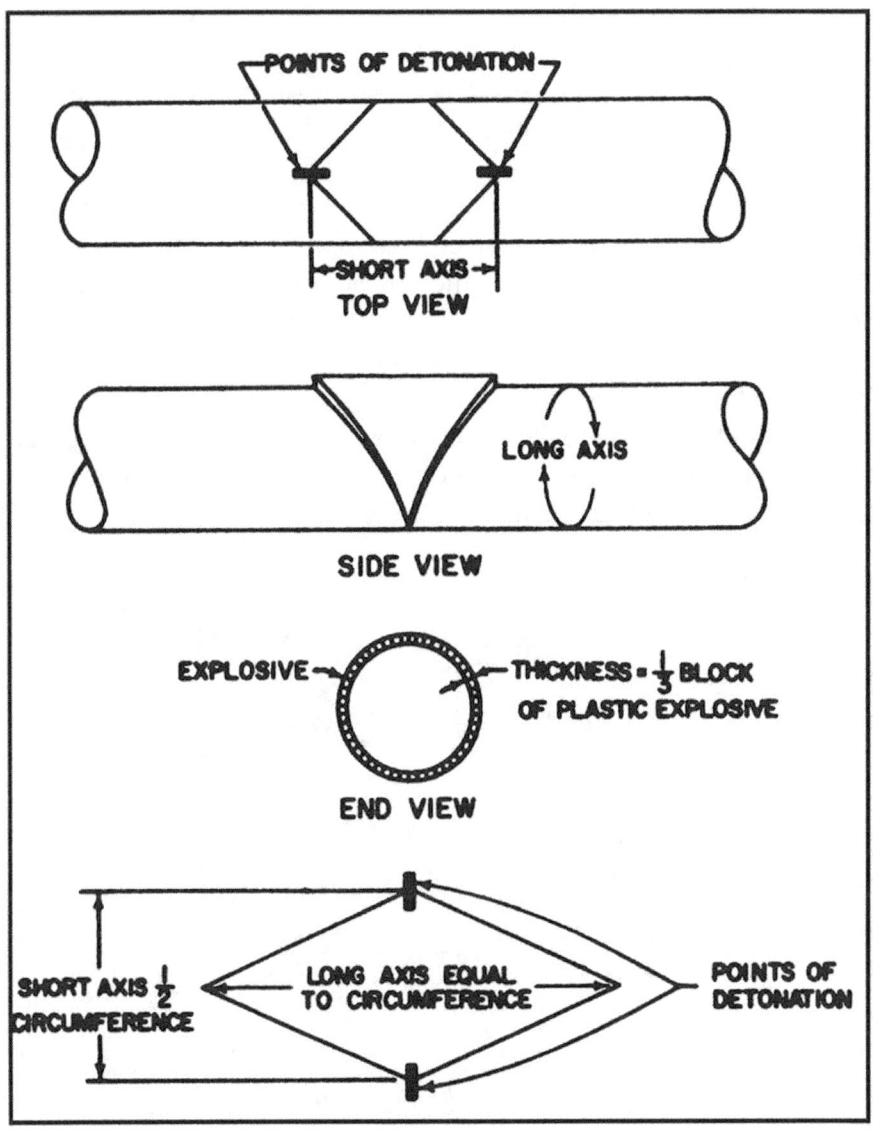

Figure 19. Diamond Charge

The diamond charge is more time consuming to construct, and requires both more care and more materials to prime. Transferring the charge dimensions to a template or cardboard or even cloth permits relatively easy charge construction (working directly on the target is extremely difficult). The completed wrapped charge is then transferred to the target and taped or tied in place, insuring that maximum close contact is achieved. The template technique should be used for both the saddle and diamond charges.

Ribbon Charge. To out flat or non-cylindrical steel targets the ribbon charges produce excellent results at a considerable savings in explosive. Dimensions are as follows: the thickness of the charge is equal to the thickness of the target to be cut. (Note: never construct a charge less than 1/2 inch thick.) Width of the ribbon is equal to twice the thickness of the target. Length of the charge is equal to the length of the desired cut. Prime from an end; and for relatively thin charges, build up the end to be primed. Build up corners if the charge is designed to cut a target such as an I-beam. Tamping is unnecessary with the ribbon charge. A frame can be constructed out of stiff cardboard or plywood to give rigidity to the charge and to facilitate handling, carrying, and emplacing it. The ribbon charge is effective only against targets up to 2 inches thick, which effectively accounts for the great majority of flat steel targets likely to be encountered.

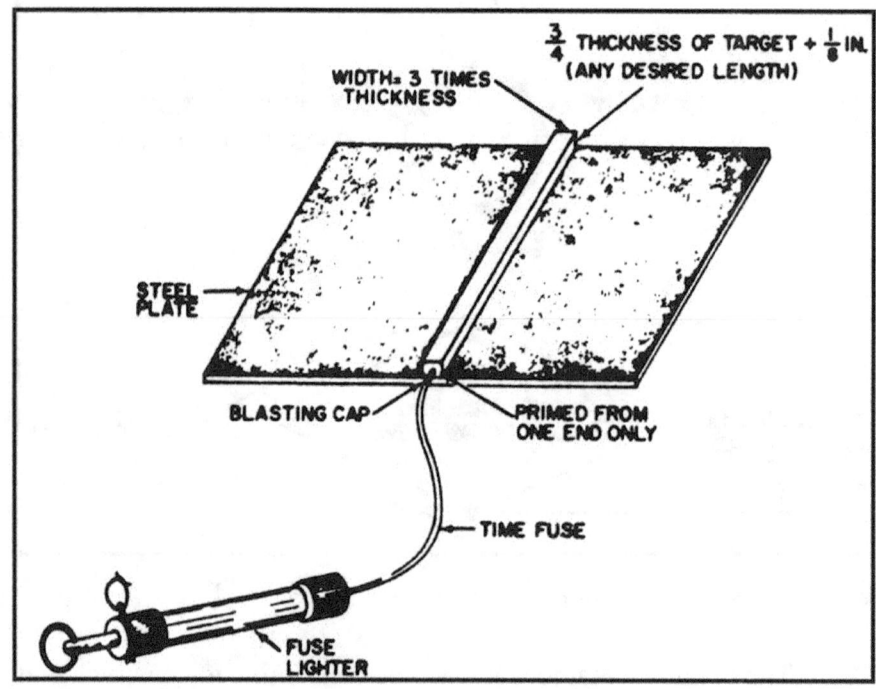

Figure 20. Ribbon Charge

Paste Explosive. Excellent results have been obtained in cutting railroad rails and other steel targets by using improvised paste explosive. An example of paste effectiveness follows: the standard steel cutting formula, $P = 3/8\ A$, yields an answer of 560 grams of explosive required to cut a rail 90 lb/yard. Eighty grams of paste explosive were actually used, and this charge removed more than a foot of the track.

Shaped Charges. If available, manufactured shaped charges will always give results far superior to those produced by any improvised shaped charges. The angle of the cavity of an improvised shaped charge should be between 30 and 60 degrees. Stand-off should be from 1 to 2 times the diameter of the cone. Height of the explosive, measured from the base of the cone should be twice the height of the cone. Exact center priming and tightly packed C4 is important. Trial and error experimentation in determining optimum stand-off is necessary. A point worth mentioning in preparing hollow-bottomed bottles for shaped charge use is to hold the bottle upright when burning the string soaked with gasoline. As the flame goes out submerge the bottle, neck first, in water; and if properly done, the bottle will break cleanly where the string was burned. Hemispherical cavities will produce more surface damage on the target but less penetration. A true cone with an angle of approximately 45 degrees will produce more penetration, which ultimately is the desired results. (See figure 21.)

Platter Charge. The platter charge has been developed to breach volatile fuel containers and ignite their contents, from distances up to 50 yards depending on the size of the target. The platter can also employed to destroy small electrical transformers or other similarly "soft" targets, again from a distance.

Platters do not have to be round or conclave although a round, concave platter is undoubtedly best. (The concave side of the platter faces the target, and the explosive goes on the reverse, or convex, side.) First, square or rectangular platters are permissible with steel being the best material. Platter size preferably should be between 2 to 6 pounds, and weight of explosive should approximately equal platter weight. The explosive should be uniformly packed behind the platter and it must be primed from exact rear center. (Build up the C4 in the center of the charge if necessary to insure detonation.) A container is completely unnecessary for the platter charge as long as some way is found to hold the plastic firmly to the platter; tape is acceptable. The range is something in the neighborhood of 25 to 50 yards. With practice, a good demolitionist can hit a 55 gallon drum, a relatively small target, at 25 yards 90 percent of the time. The largest glass or ceramic platters do not give results approaching those of steel.

Improvised Claymore or Improvised Grapeshot Charge. One of the most effective antipersonnel charge that can be improvised in the field requires the use of C4 and only a few other widely available materials. A container such as a number 10 can is excellent, although virtually any sized can or container could obviously be used. The ratio of projectiles ideally should be small pieces of steel although other objects can be used. Iron, brass, and stones can be used but, for the more fragile items, reduce the weight of explosive and add a few inches of buffer material, either earth or leaves, between the explosive and the projectiles. To prepare the charge, place the projectiles in the container. Next place a layer of thick cloth, felt, cardboard, wood, or some silmilar material over the projectiles. Whenever in doubt about the amount of explosive to use, use a lighter rather than a heavier charge. Again trial and error experimentation is extremely important in arriving at the best charge loading. The effectiveness of the finished product in this case makes all such efforts extremely worthwhile. Pack the C4 uniformly behind the separator disc, prime from exact rear center, and aim the charge toward the center of the desired target area. We obtain excellent results, in dispersion, penetration, and range, by using expended .45 caliber slugs. The main problem to guard against is the tendency to overcharge. A relatively small amount of C4 is all that is necessary to propel the projectiles: anything more will pulverize them.

Soap Dish. An excellent charge for both rupturing and igniting the contents of volatile fuel containers is the soap dish which, in contrast to the platter charge, must be placed directly on the target. Using a standard GI soap dish, containers up to 100 gallons can be reliably attacked. Charge proportions are as follows: Equal parts by volume of plastic explosive and thermate mix are placed in the container to be used, always insuring that the incendiary mix is placed against the target. The mix can be composed of a number of compounds, among which are: three parts potassium chlorate and two parts sugar, or two parts aluminum powder to three parts of ferric oxide. In lieu of these improvised incendiary mixes, the contents from thermate grenades can be used or military dynamite may be used as the explosive and match heads as the incendiary. As a rule of thumb, a thin cigar box (from an inch and a half to an inch and three-quarters thick) loaded as specified above with one-half C4 to one-half incendiary mix will reliably rupture and ignite volatile fuel containers of up to 1,000 gallon capacity. A charge of approximately twice the size will successfully attack containers of up to 5,000 gallon capacity. To prime the charges always insure that the cap is inserted into the explosive and not the incendiary mix. Holding the charge in place may be accomplished by the use of magnets or adhesive. Always insure that the charge is placed below the fuel level in the container.

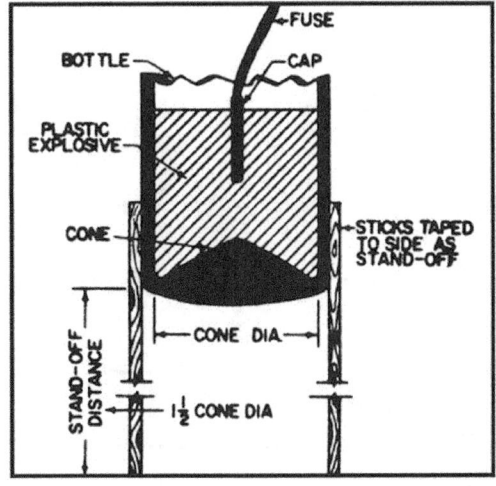

1-STANDOFF-1 to 2 times diameter of cone
2-CONE ANGLE-30° to 60°
3-EXPLOSIVE DEPTH-2 times height of cone
4-DETONATED REAR DEAD CENTER

Figure 21. Shaped Charge.

Opposed Charge. (Also called the "counterforce" or "ear muff " charge.) Within its limitations, which are quite restrictive, the opposed charge offers dramatic savings in explosives for destroying reinforced concrete targets. The rule of thumb for construction is as follows: for each foot of target thickness, up to a maximum of 4 feet, use 1 pound of C4; for fractions of a foot, go to the next higher pound. Divide the total amount of C4 exactly in half, placing one half of the charge on each side of the target, diametrically opposite each other. (This brings up one limitation, the requirement to have two sides of the target accessible.) Prime the two charges to detonate exactly simultaneously, and the target will be destroyed as the shook waves meet in the center of the target and, in effect, cause it to virtually explode from within. It will be noted that the charge size has been reduced by one-half the amount called for in previous publications. This charge is onyl effective and reliable against targets that are approximately square and not much much larger than 4 feet square. (See figure 22.)

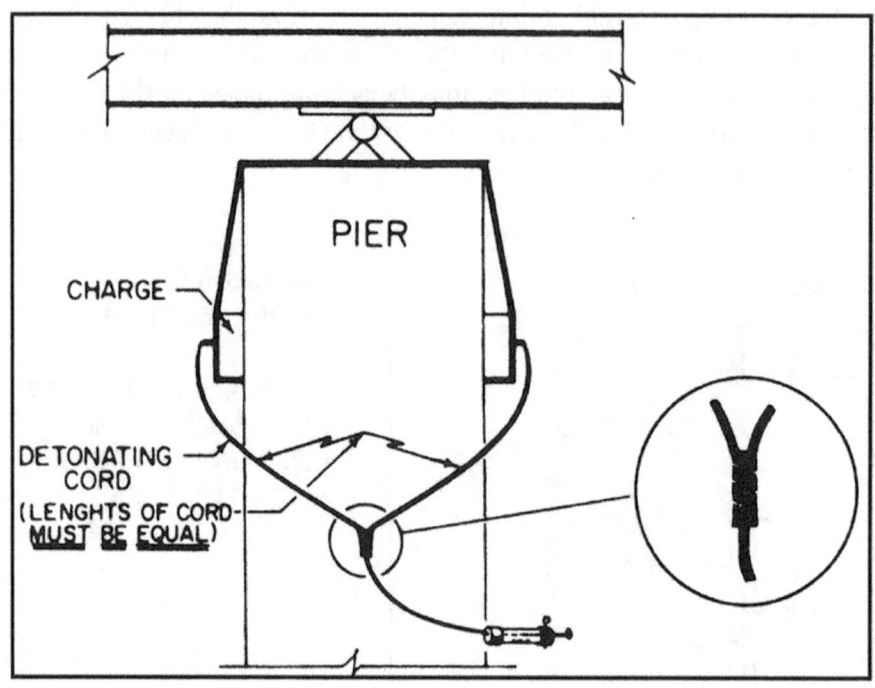

Figure 22. Opposed Charge.

Improvised Cratering Charge. Ammonium nitrate fertilizer is a material that is readily available in many parts of the world. With AN and one other simple ingredient we have the ability to "tailor make" cratering charges to practically any size or configuration. A rule of thumb for the construction of an improvised cratering charge is as follows: to each 25 pounds of ammonium nitrate fertilizer, which should be the prilled or pelleted variety, add approximately 1 quart of diesel fuel, motor oil, or gasoline. The motor oil may be drained from a crankcase, which will not impair the effectiveness of the charge. Allow the charge to soak for 1 hour, prime with 1 pound of TNT, or its equivalent, tamp well in an appropriate bore hole, and detonate. The results obtainable with this charge compare very favorably with the manufactured variety. The prilled Ammonium Nitrate fertilizer should be of a kind having at least 33 1/3 percent nitrogen content and care should be exercised to see that the fertilizer used is not damp. Obviously it cannot be left for extended times in a borehole or water will reduce the effectiveness of the charge. When difficulty is encountered in producing a borehole diameter that is capable of accommodating the bulk of the manufactured 40 pound cratering charge, 8 1/4 by 17 inches, excellent results can be obtained by pouring and tamping the improvised AN cratering charge into the available space.

Improvised Ammonium Nitrate Satchel Charge. While the cratering charge referred to earlier is undeniably a good one, it is really only suitable for cratering use. A more manageable charge can be produced from AN, using wax as the second ingredient, rather than oil. The procedure for making this charge is merely to melt ordinary parafin and stir in AN pellets, insuring that the wax is thoroughly mixed with the AN while still hot. Before the mixture hardens add a one-half pound block of TNT, or its equivalent, as a primer. A number 10 can makes a good container for this charge although practically anything may be used. The addition of suitable shrapnel material and a handle to the exterior of the charge makes an excellent expedient satchel charge that is more manageable than the AN and oil mixture and much less susceptible to moisture. In fact, this charge can be stored for extended periods without regard to humidity and without loss of effectiveness.

Dust Initiator. The employment of a small initiator charge to make use of explosive energy provided at a target site is an economical means of destroying certain types of targets. An improvised dust initiator charge can be constructed as follows: To make the standard 1 pound charge use half explosive and half incendiary mix. The explosive may be either powdered TNT (obtained by crushing the TNT in a canvas bag) or C3. C4 does not properly mix with the incendiary and will not produce the desired result. The incendiary mix may be two parts of aluminum powder to three parts of ferric oxide; magnesium powder may be used in lieu of aluminum powder. If used with powdered TNT, the two should be thoroughly mixed. If used with C3, the incendiary mix should be thoroughly mixed throughout the half-pound of explosive. The dust initiator requires a "surround" which is merely the addition of a suitable, finely divided (dust) material or a volatile fuel such as gasoline. The DI works best in an inclosed space; and such targets as boxcars, warehouses, and other relatively windowless structures are best suited to an attack by this means. A rule of thumb for its employment is that three to five pounds of cover or surround should be provided for each 1,000 cubic feet of target. The 1 pound DI charge will effectively disperse and detonate up to 40 pounds of cover charge. The effect of the surround as it is first scattered and then detonated by the long-lasting flame of the DI's explosion is to increase the internal

43

explosive pressure from 500 to 900 percent over the effect of the DI being detonated without a surround. If used with gasoline the optimum results are obtained by only using 3 gallons of the fuel. The addition of more gasoline not only does not produce better results, the fuel usually will not even be detonated. A large number of dust materials can be used as a surround, including coal dust, cocoa, bulk powdered coffee, confectioners sugar, tapioca, and powdered soap. A good expedient DI charge can also be produced by packaging the contents of two thermate grenades around a stick of military dynamite. (Note that this is just the DI charge to which a surround must be added.)   Figure 18.

VI.   IMPROVISED INCENDIARIES, EXPLOSIVES AND DELAY DEVICES.

Caution: As a general rule improvised explosives and incendiaries are much more dangerous to handle than conventional explosives. Such mixtures as the chlorate-sugar mix mentioned below can be ignited or detonated by a single spark, excessive heat, or merely by the friction generated by stirring or mixing the ingredients together. The danger in handling these items cannot be over emphasized.

Chlorate-Sugar Mix. This mixture can be either an incendiary or an explosive. Sugar is the common granulated household variety. Either potassium chlorate or sodium chlorate may be used; potassium is preferred. Proportions can be equal parts by volume, or 3 parts of chlorate to 2 parts of sugar preferred. Mix in or on a non-sparking surface. Unconfined, the mix is an incendiary. Confined in a tightly capped length of pipe it will explode when a spark is introduced. Such a pipe bomb will produce casualties, but will not be suitable for breaching or cutting tasks. Concentrated sulfuric acid will ignite this fast burning incendiary mixture. Placing the acid in a gelatin capsule, balloon, or other suitable container will provide a delay (length of which depends on how long it takes the acid to eat through the container).

Potassium Permanganate and Sugar. Another fast burning, first fire mix is obtained by mixing potassium permanganate, 9 parts, to one part sugar. It is somewhat hotter than the chlorate sugar mix, and can be ignited by the addition of a few drops of glycerine.

Sawdust and Wax. An effective and long burning incendiary can be produced by adding molten wax or tar to sawdust. The advantage of this incendiary is that its components are truly universally available.

Matchhead. A quantity of matchheads cut from common safety matches will make either a fast burning incendiary or, if confined, an explosive. A length of pipe filled with matchheads and capped and fuzed makes an effective antipersonnel bomb. Again extreme caution must be exercised in handling of matchheads in bulk—a single spark will detonate or ignite them.

Improvised Napalm. To either gasoline or kerosene add finely cut soap chips. Pure SOAP must be used, not detergents. Working in the open, use a double boiler with the bottom portion filled approx 3/4 full of water. Heat until fuel comes to a boil and then simmers. Stir constantly until the desired consistency is reached: Remember that it will thicken further on cooling. Trial and error experimentation will determine proper amounts for best results.

Improvised Thermite Grenade. The main burning agent, the thermite, is composed of 3 parts of iron oxide to 2 parts of aluminum powder. A ceramic flower pot makes a good container for the thermite. A potassium chlorate and sugar first fire mix of 3 parts chlorate and 2 parts sugar is placed in a paper tube running down through the thermite. When the chlorate is ignited, it in turn ignites the thermite, which can be used to attack mild steel. This thermite mix burns at approx. 4,000 degrees.

Molded Brick Incendiaries. Proportions are 3 parts aluminum powder, 4 parts water and 5 parts plaster of parts. Mix the aluminum and plaster thoroughly together, then add the water and stir vigorously. Pour the resulting mix into a mold, let harden, and dry for 2 to 3 weeks. While they are difficult to ignite, a dry mix of 3 parts of oxide and 2 parts of alumimum powder should be used. These bricks burn with intense heat and are suitable for melting mild steel.

C4 As an Incendiary. Most plastic explosives, including C3 and C4 can be used as an incendiary. They are easy to ignite and burn with a hot flame of long duration.

Sulfuric Acid can be used to ignite chlorate and sugar. An expedient method of obtaining sulfuric acid is as follows: Drain the liquid from one or more wet cell batteries, place it in a glass, pottery or ceramic container, and heat it. As the liquid comes to a boil it will begin to emit a dense white smoke. Remove the remaining liquid from the heat, allow to cool, and place it in a tightly stoppered glass bottle. Test the acid before each operational use.

Fire bottle. Fill a glass bottle about one-fifth to one-fourth full with sulfuric acid. Fill the remainder with gasoline, kerosene, or a combination of both. Add water to potassium chlorate and sugar mix, and soak rags in the mix. Wrap the rags around the bottle, tie in place, and allow to dry. When thrown, the bottle will break, the acid will ignite the chlorate sugar saturated rags, which in turn will ignite the fuel.

VII.   MISCELLANEOUS IMPROVISED DEMOLITIONS.

Thermite. Use any size can with sticks tied or taped to sides and cut small hole in bottom. Cover bottom with paper. Place round stick wrapped in paper in middle of can. Fill bottom of can 1/4 inch with magnesium. Over this place mixture of 3 parts ferric oxide and 2 parts aluminum powder. Remove stock and fill hole with mixture 3 parts potassium chlorate and 1 part sugar. On top of this place paper bag containing chlorate-sugar mixture. Place fuze in top. tamp with dirt or clay.

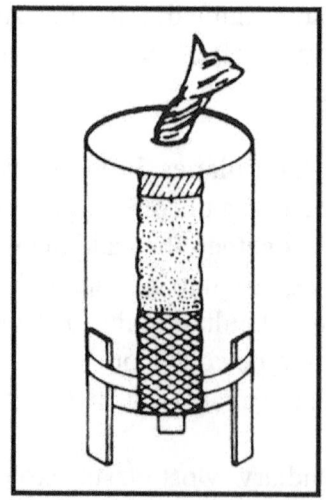

Figure 23. Thermite.

46

Molotov Cocktail. Fill bottle with napalm, jelly gas or 2 to 1 ratio mixture of gas and oil. Use wick of rag or cotton dipped in wax. Light before throwing.

Figure 24. Molotov Cocktail.

Satchel Charge. Fill #10 can with mixture of ammonium nitrate and melted wax, stirring vigorously to insure a complete mix. Prime with small amount of C4 or TNT before mixture hardens. Add a rope handle for convenient improvised satchel charge.

Improvised Black Powder.

Materials required	Percent by Wt.	Parts by Vol.
Potassium Nitrate	74	25
Powdered Charcoal	16	3
Sulpher	10	2

PROCEDURE.

(a) Dissolve potassium nitrate in water using a ratio of three parts weight of water to one part nitrate.

(b) In a second container, dry mix the powdered charcoal and sulpher by stirring with a wooden stick or rotating in a tightly closed container.

(c) Add a few drops of potassium nitrate solution to the dry mixture and blend to obtain a thoroughly wet paste. Then add the rest of the solution and stir.

(d) Pour the mixture into a shallow dish or pan and allow to stand until it evaporates to a paste-like consistency. Mix the paste thoroughly with a wooden stick to assure uniformity and set aside for further drying.

(e) When the mixture is nearly dried, granulate by forcing through a piece of wire screening. The granules are then spread thinly and allowed to dry.

e. Improvised Fuse.
String Fuse-(Hot) 3/4 cup water, 1 teaspoon potassium chlorate-boil 30 minutes.
String Fuse-soak in gasoline and dry. Burns slowly.
String Fuse-(Cold) 3/4 cup water, 2 teaspoons potassium nitrate.

f. Improvised Grenades.
7.5 parts potassium nitrate or sodium nitrate, 1.5 parts charcoal. 1 part sulphur (no detonator, just fuse)
3 parts sodium chlorate, 2 parts sugar. Contain in a lead pipe (no detonator. just fuse)

g. Types of Delays:

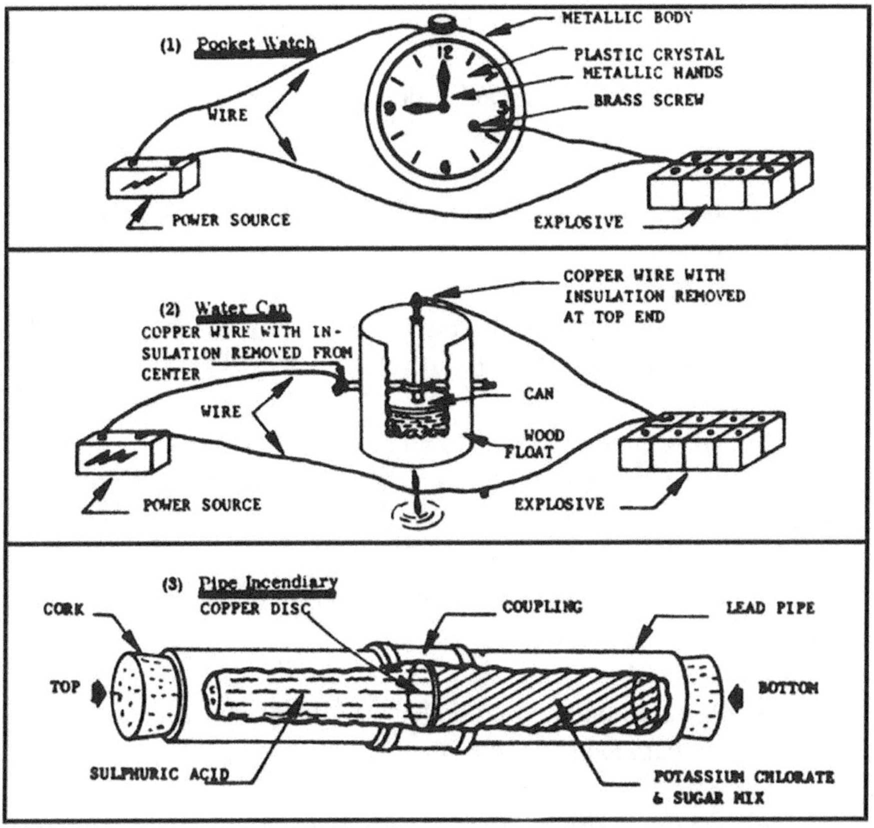

METALLIC BODY

(1) Pocket Watch

PLASTIC CRYSTAL
METALLIC HANDS

BRASS SCREW

WIRE

POWER SOURCE

EXPLOSIVE

COPPER WIRE WITH
INSULATION REMOVED
AT TOP END

(2) Water Can
COPPER WIRE WITH IN-
SULATION REMOVED FROM
CENTER

CAN

WIRE

WOOD
FLOAT

POWER SOURCE

EXPLOSIVE

(3) Pipe Incendiary
COPPER DISC

CORK

COUPLING

LEAD PIPE

TOP

BOTTOM

SULPHURIC ACID

POTASSIUM CHLORATE
& SUGAR MIX

Figure 25. Types of Delays

h. Flame Illuminator. Fill container 3 inches from top w/thickened fuel and seal tightly. Put three wraps dot cord on top onside of rim pack with dirt or mud. Wrap grenade with det cord and place next to container. Tie to main det cord line.

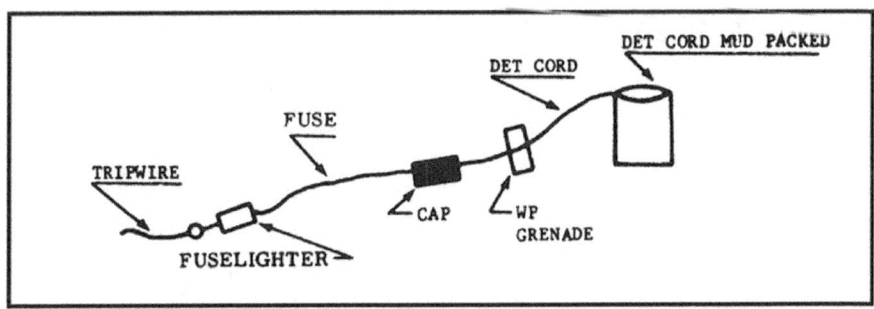

DET CORD MUD PACKED

DET CORD

FUSE

TRIPWIRE

CAP

WP
GRENADE

FUSELIGHTER

Figure 26. Flame Illuminator.

49

i. Husch Flare (Burns for 90 minutes, lights dia. 50 meters).

Remove cross bars from metal 60 mm mortar can.

Punch 3 3/8" holes in each side 1/2 way between top and bottom.

Punch hole not bigger than 1/8" in bottom of 81mm mortar metal shell container.

Temporarily fill holes, fill container 3/4 full w/thickened fuel, apply heavy grease to caps and affix tightly.

Place 81 containers caps down in 60 mortar container, wedge tight with stones, etc, then fill 60 mm mortar can with thickened fuel up to holes.

Remove plugs from 1/8" holes in bottom of 81mm shell container.

Tie illumination hand grenade between 81 mm cans just above level of 60 mm can. Run trip wire from grenade pin.

j. Dried Seed Timer.

DRIED SEED TIMER

A time delay device for electrical firing circuits can be made using the principle of expansion of dried seeds.

MATERIAL REQUIRED:

Dried peas, beans or other dehydrated seeds
Wide- mouth glass jar with nonmetal cap
Two screws or bolts
Thin metal plate
Hand drill
Screwdriver

50

PROCEDURE:

Determine the rate of rise of the dried seeds selected. This is necessary to determine delay time of the timer.

Place a sample of the dried seeds in the jar and cover with water.

Measure the time it takes for the seeds to rise a given height. Most dried seeds increase 50% in 1 to 2 hours.

Cut a disc from thin metal plate. Disc should fit loosely inside the jar.

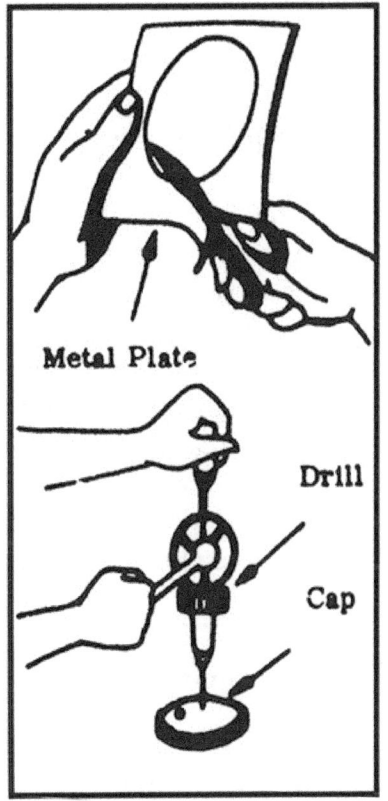

Metal Plate

Drill

Cap

NOTE:
If metal is painted, rusty or otherwise coated, it must be scraped or sanded to obtain a clean metal surface.

Drill two holes in the cap of the jar about 2 inches apart. Diameter of holes should be such that screws or bolts will thread tightly into them. If the jar has a metal cap or no cap, a piece of wood of plastic (NOT METAL) can be used as a cover.

Turn the two screws or bolts through the holes in the cap. Bolts should extend about one in. (2 1/2 cm) into the jar.
IMPORTANT: Both bolts must extend the same distance below the container cover

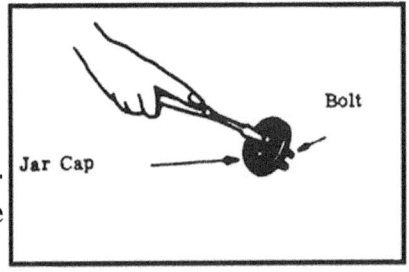

Bolt

Jar Cap

Pour dried seeds into the container. The level will depend upon the previously measured rise time and the desired delay.

Place the metal disc in the jar on top of the seeds.

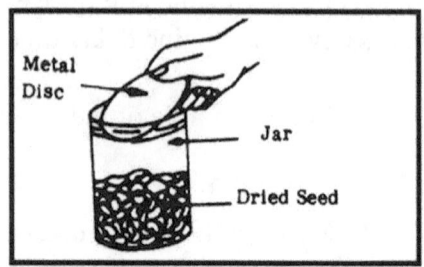

HOW TO USE:

Add just enough water to completely cover the seeds and place the cap on the jar.

Attach connecting wires from the firing circuit to the two screws on the cap.

Expansion of the seeds will raise the metal disc until it contacts the screws and closes the circuit.

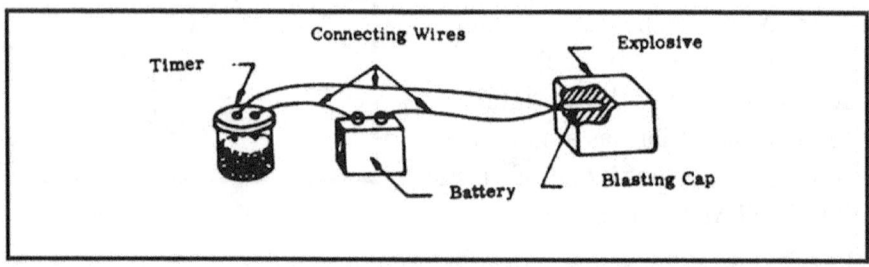

k. Tin Can Grenade.

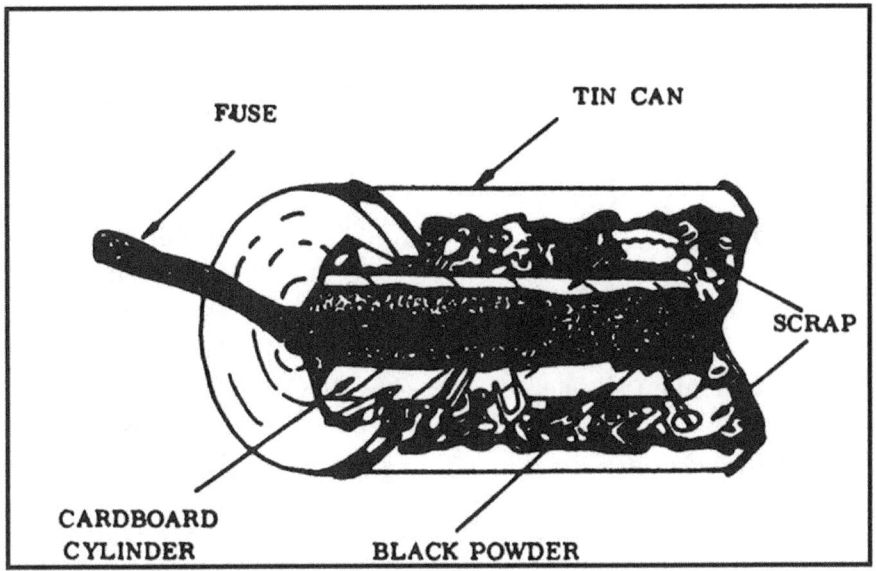

Figure 33. Tin Can Grenade.

MATERIALS REQUIRED.

Tin can, jar or similar container.
Bolts, nuts, metal scrap, etc.
Commercial or improvised Black Powder.
Commercial or improvised fuse cord.
Cardboard or heavy paper and tape.

PROCEDURE.

Tape cardboard or heavy paper into a cylinder approximately one half the diameter of the tin can or other container.

Insert the fuse into one end of this cylinder, pack tightly with black powder and tape the ends closed.

Insert the cylinder into the can as shown in Figure 33 and surround with bolts, nuts, metal scrap and/or stones. Close the can with a lid which has a hole in the center for the fuse to pass through. If the container used has no lid, it may be closed with a piece of wood, metal or cardboard of the required size taped in place.

53

VIII. CHEMICALS

Chemical	Symbol	Source
Potassium Permanganate	KMN04	Drug Store, Hospital, Gym
Potassium Chlorate	KCLO3	Drug Store, Hospital, Gym
Potassium Nitrate	KNO3	Fertilizer, Explosive Mgfr
Sodium Nitrate	NaNO3	Fertilizer, Glass Mgfr
Ammonium Nitrate	(NH4)NO3	Fertilizer, Explosive Mgfr
Ferric Oxide	Fe2O3	Hardware or paint store
Powdered Aluminum	AL	Paint store, electric, auto
Magnesium	Mg	Auto Mgfr, Machine, Chemical
Glycerine	C3H5(OH)3	Drug Store, Soap, Candle Mgfr
Sulphuric Acid	H2SO4	Garage, Machine Shop, Hospital
Sodium Chlorate	NaClO3	Match, Explosive Mgfr, Plant
Sulphur	S	Drug Store, Match Mgfr

IX. DELAYS.

Cigarette (in match book or box)
Candle (surrounded by inflammable material)
Spark (from short circuited electrical wires)

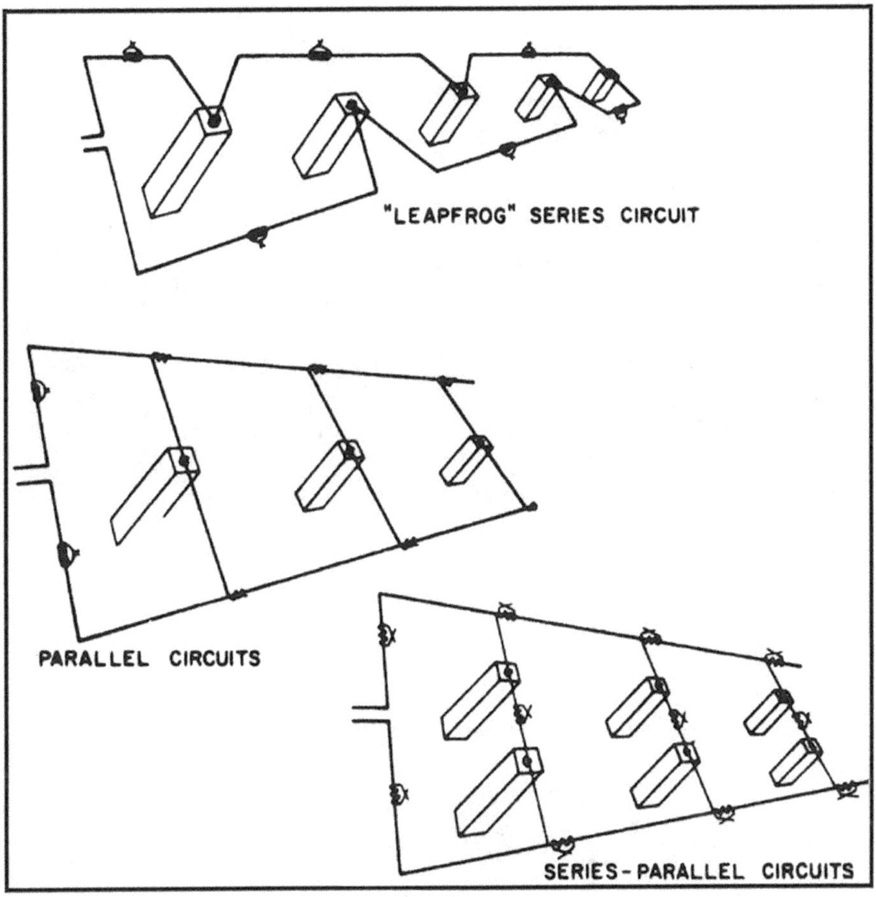

X. DEMOLITION DATA.

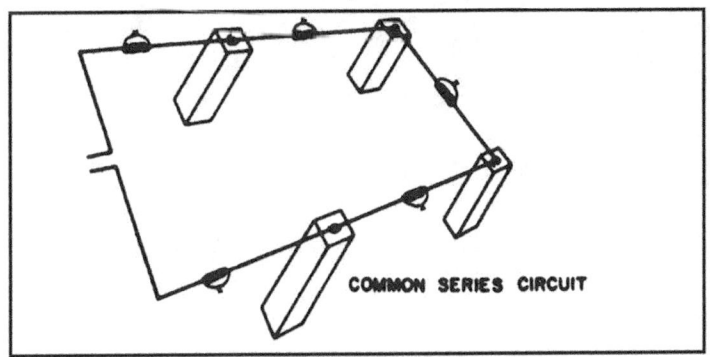

"LEAPFROG" SERIES CIRCUIT

PARALLEL CIRCUITS

SERIES-PARALLEL CIRCUITS

Figure 34.   Leapfrog series circuit.
Parallel circuits.
Series-parallel circuits.

COMMON SERIES CIRCUIT

Figure 35. Common Series Circuit.

Cratering

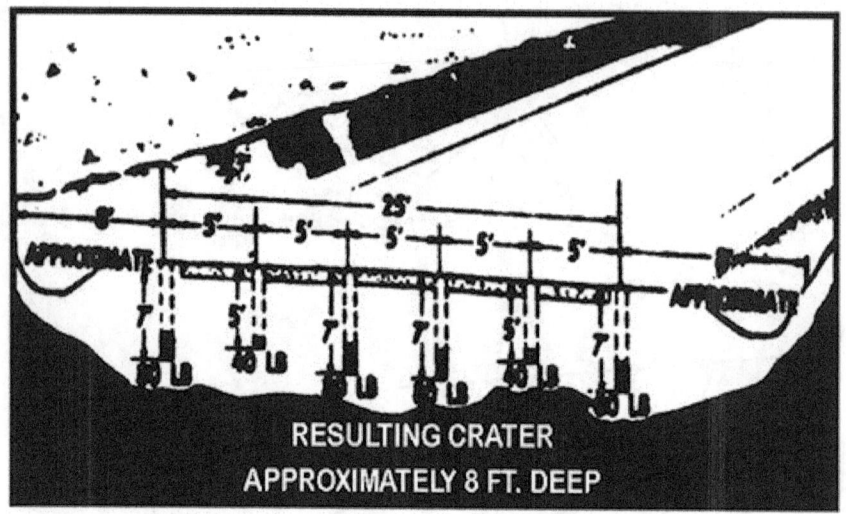

Figure 36. Placement of charges for deliberate road crater.

Figure 37. Placement of charges for hasty road crater.

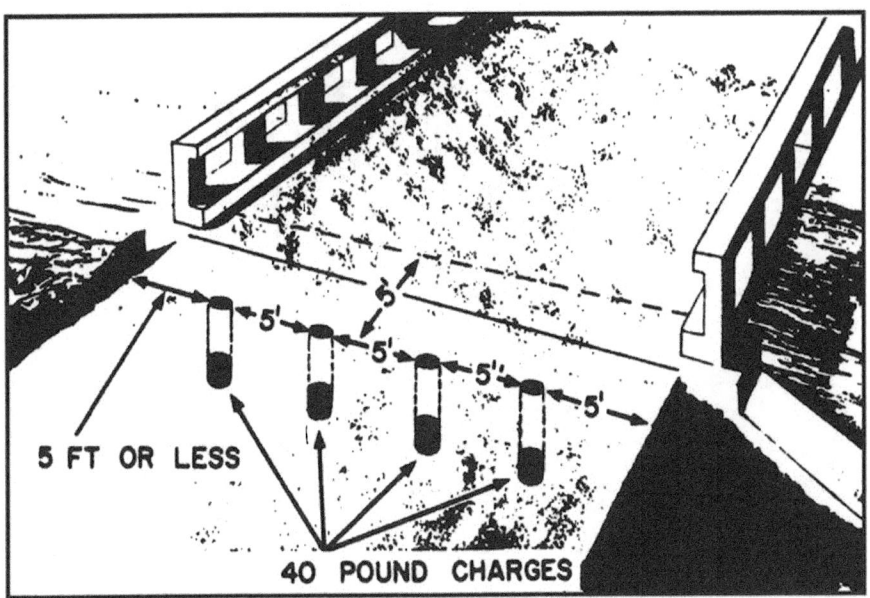

Figure 38. Charges placed in fill behind reinforced concrete abutment 5 feet or less in thickness. (The 5-5-5-40 method.)

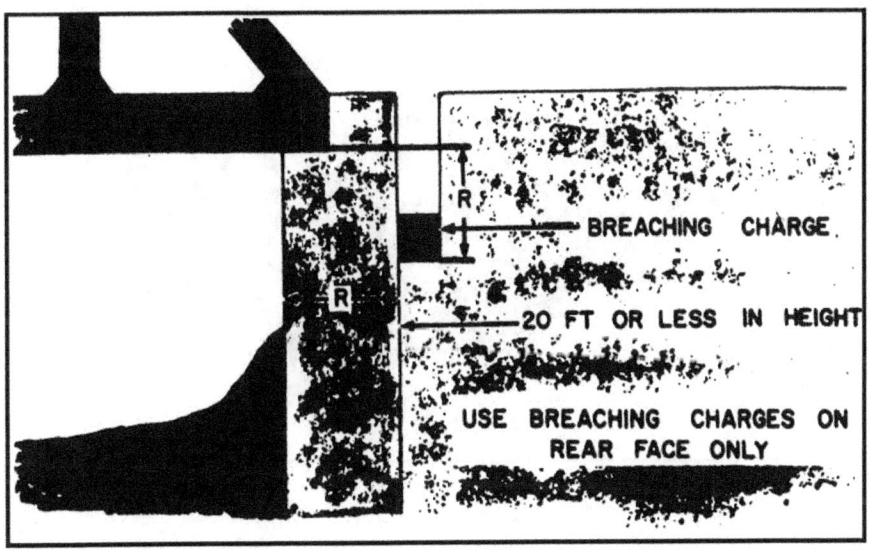

Figure 39. Placement of charges behind concrete abutment more than 5 feet thick.

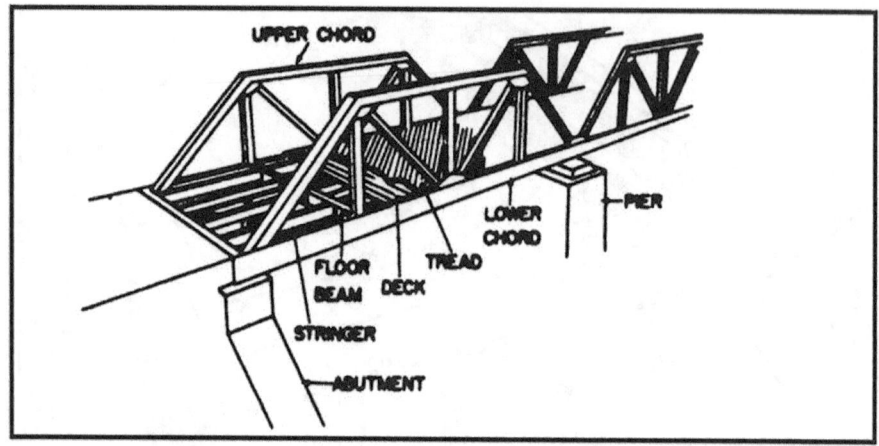

Figure 40. Parts of fixed bridge.

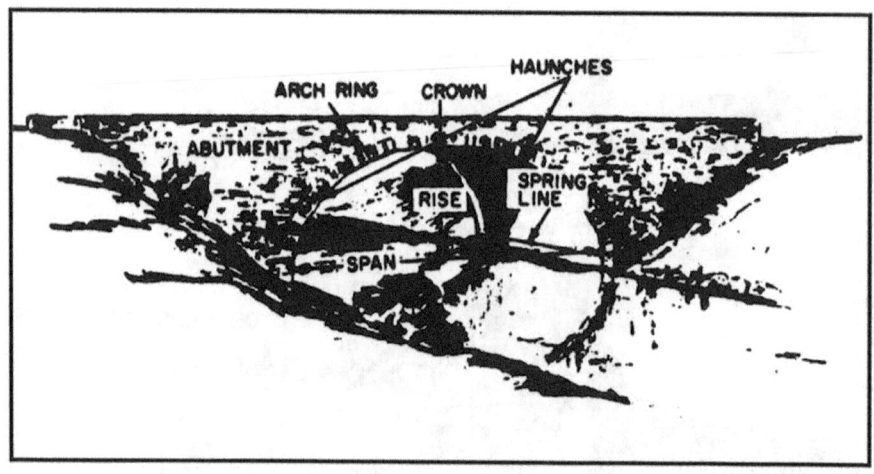

Figure 41. Arch components.

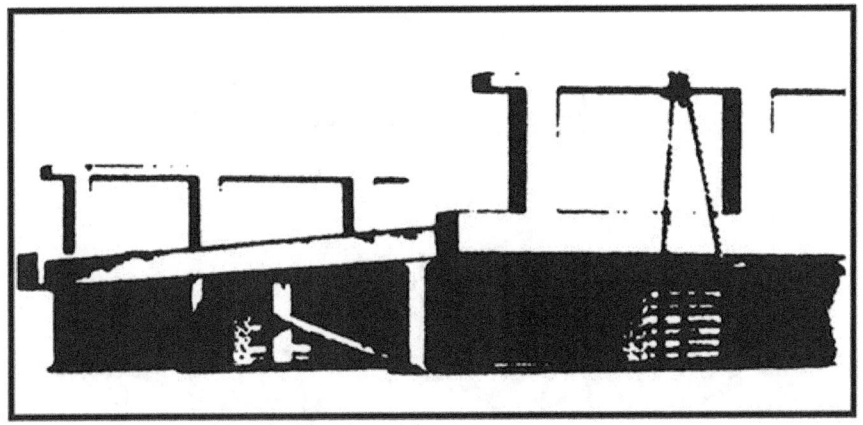

Figure 42. Steel stringer bridge.

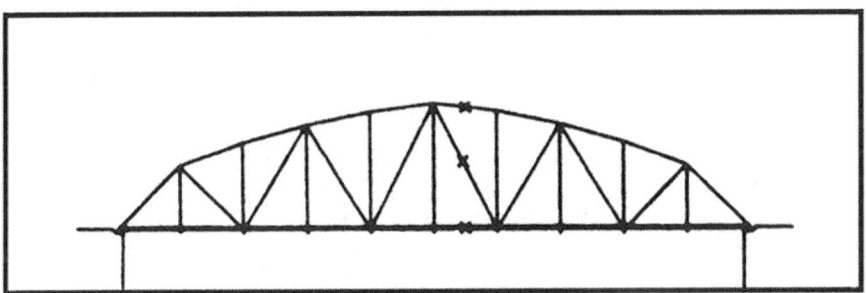

Figure 43. To cut upper and lower chords.

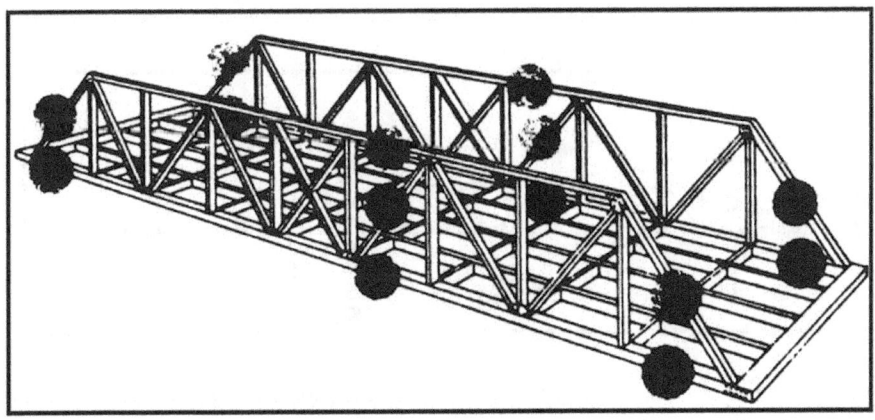

Figure 44. To cut trusses into segments.

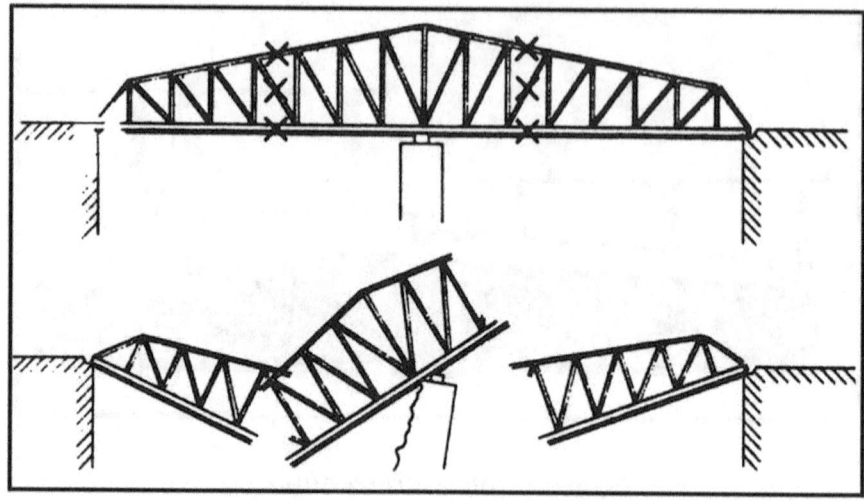

Figure 45. Continuous span truss.

Figure 46. Cantilever truss with suspended span.

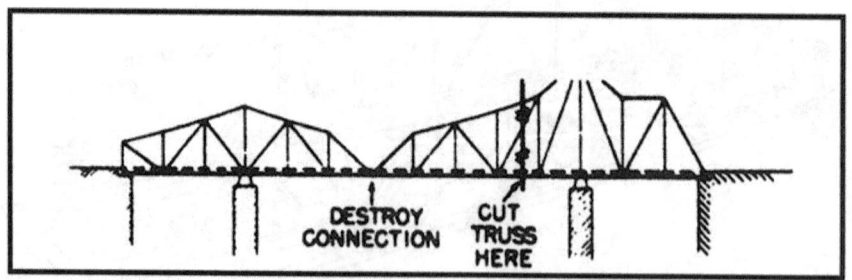

Figure 47. Cantilever truss without suspended span.

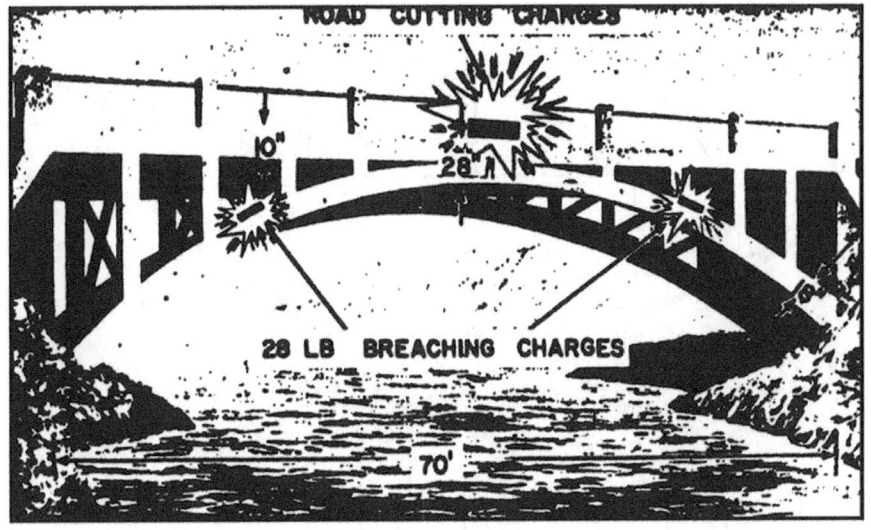

Figure 48. Reinforced concrete open spandrel arch bridge.

Figure 49. Filled spandrel arch bridge.

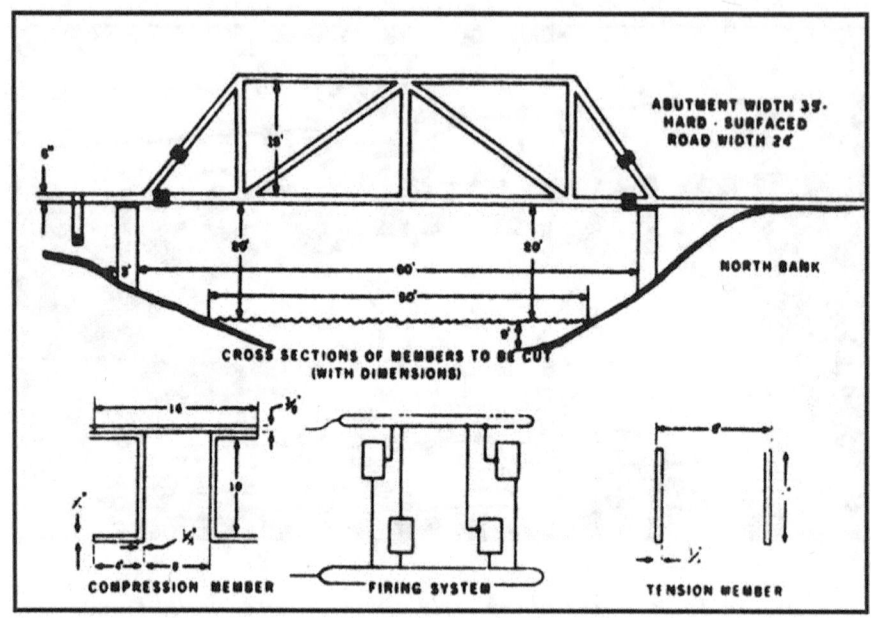

Figure 40.

XI.   TARGET RECONNAISSANCE REPORT GUIDE.

Target: _____

Location: _____

Times Observed: _____

General Description: _____

Proposed Action: _____

Route: To and from target areas, approach and withdrawal routes, rallying points, mission support sites, cache sites and final assembly areas may be selected.

Requirements: (Determine availability before recon)

Explosives: _____

Equipment: _____

Personnel: _____

Time: _____

Remarks:

Unusual features of site: _____

Guard system: _____

Labor and time estimate required for bypass or repair:

Sketches: (On reverse side)

Situation map sketch (overhead view):

Magnetic north,

Principal terrain concealment,

Avenues of approach to target,

Direction of enemy, etc.

Line drawing of target (side or angle view):

Critical over-all dimensions and placement of charges.

Cross sections of members to be cut (cut-away view):

Exact dimensions.

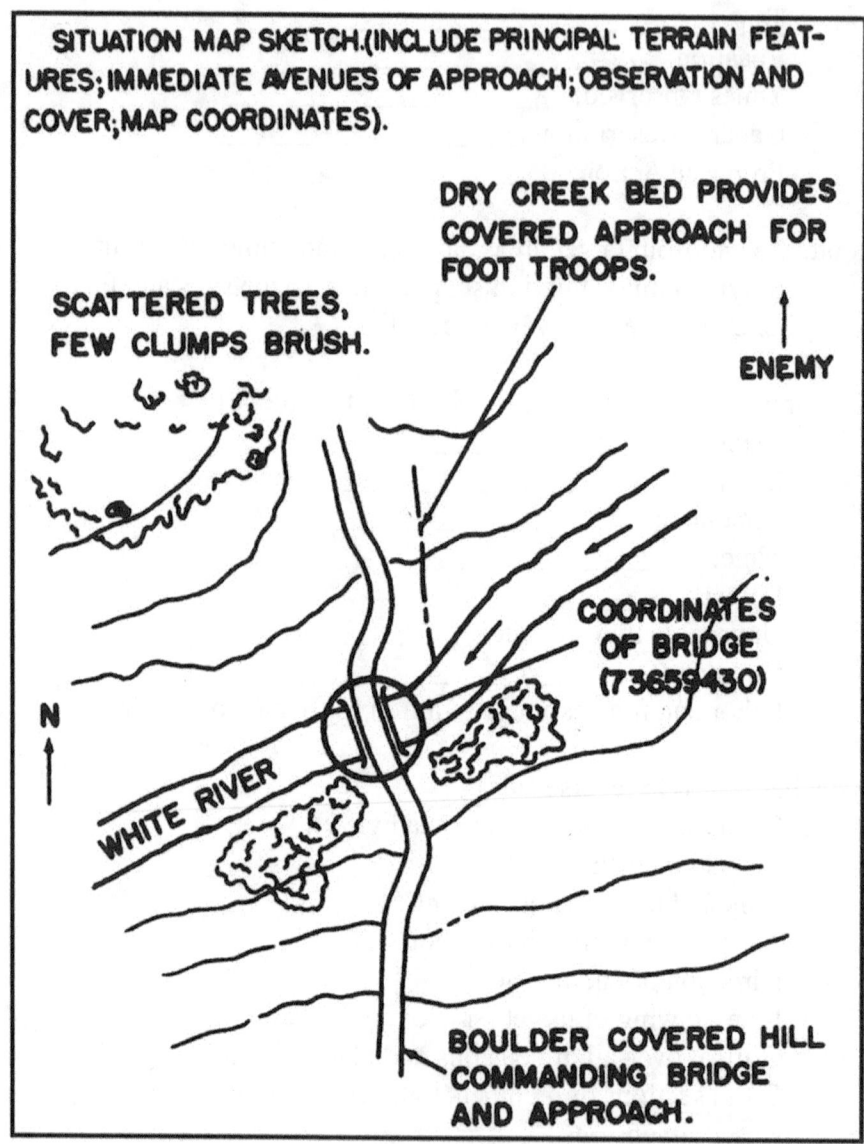

Figure 51.

CHAPTER 4
AIR OPERATIONS

I. PREPLANNED AIR RESUPPLY OPERATIONS:

Automatic Resupply Plan. This plan provides for initial automatic replacement of essential equipment and supplies, primarily communications equipment, immediately after infiltration. Preinfiltration planning includes: DZ selection, DZ markings, drop time and date, and supplies to be dropped. Immediately after infiltration provide for replacement of essential equipment and supplies, particularly communications equipment. The automatic resupply plan may be received as planned, modified, or may be cancelled after infiltration, once contact is established with the SFOB. If the detachment fails to contact the SFOB after infiltration, the drop is executed as preplanned.

Emergency Resupply Plan. This plan provides for emergency replacement of supplies and equipment essential to individual survival, communications, and combat throughout the time that the detachment is in the operational area. Preinfiltration planning includes: provisional DZ selection to be confirmed after infiltration, DZ markings, drop date and time based upon the emergency, and supplies to be dropped. After infiltration is completed and communications established with the SFOB, the emergency DZ location (which is known only to the special forces detachment members) is either confirmed or a new location is designated. The preplanned emergency resupply drop is normally executed after the detachment misses a specified, consecutive number of scheduled communications contacts.

II. DROP ZONES:

General. The selection of a DZ must satisfy the requirements of both the aircrew and the reception committee. The aircrew must be able to locate and identify the DZ. The reception committee selects a site that is accessible, reasonably secure, and permits safe delivery of incoming personnel and/or supplies.

Air considerations. Desirable terrain features. The general area surrounding the site must be relatively free from obstacles which may interfere with safe flight. Flat or rolling terrain is desirable; however, in mountainous or hilly country, sites selected at higher elevations such as level plateaus can be used. Small valleys or pockets completely surrounded by hills are difficult to locate and should not normally be used. In order to afford the air support unit flexibility in selecting the IP, it is desirable that the aircraft be able to approach the target site from any direction. There should be an open approach quadrant of at least 90° to allow the aircrew a choice when determining their approach track from the IP. DZs having a single clear line of approach are acceptable for medium aircraft, provided there is a level turning radius of 3 miles, (5 kilometers) on each side of the site (1 mile or 1.5 kilometers for light aircraft) (Figure 52).

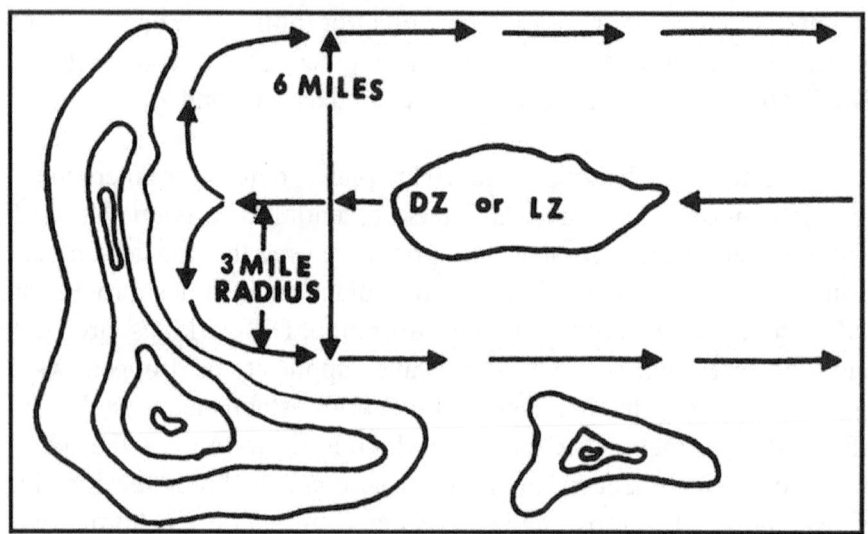

Figure 52.

Rising ground or hills of more than 1,000 feet (305 meters) elevation above the surface of the site should normally be at least 10 miles from DZ for night operations. In exceptionally mountainous areas deviations from this requirement may be made. Any deviation will be noted in the DZ report. Deviations from the aforementioned minimum distances cause the aircraft to fly at higher than desirable altitudes when executing the drop. Weather in drop areas. The prevailing weather conditions in the area must be considered. Ground fogs, mists, haze, smoke, and lowhanging cloud conditions may interfere with visual signals and DZ markings. Excessive winds also hinder operations.

Obstacles. Due to the low altitudes at which operational drops are conducted, consideration must be given to navigational obstacles in excess of 300 feet (90 meters) above the level of the DZ and within a radius of 5 miles (8 kilometers). If such obstacles exist and are not shown on the issued maps, they must be reported. Enemy air defenses. Drop sites should be located so as to preclude the aircraft flying over or near enemy air installations when making the final approach to the DZ.

Ground Considerations.
Shape and size.

The most desirable shape for a DZ is square or round. This permits a wider choice of aircraft approach directions than is normally the case with rectangular-shaped sites.

The required length of a DZ depends primarily on the number of units to be dropped and the length of their dispersion pattern.

Dispersion occurs when two or more personnel or containers are released consecutively from an aircraft in flight. The long axis of the landing pattern is usually parallel to the direction of flight (Figure 53).

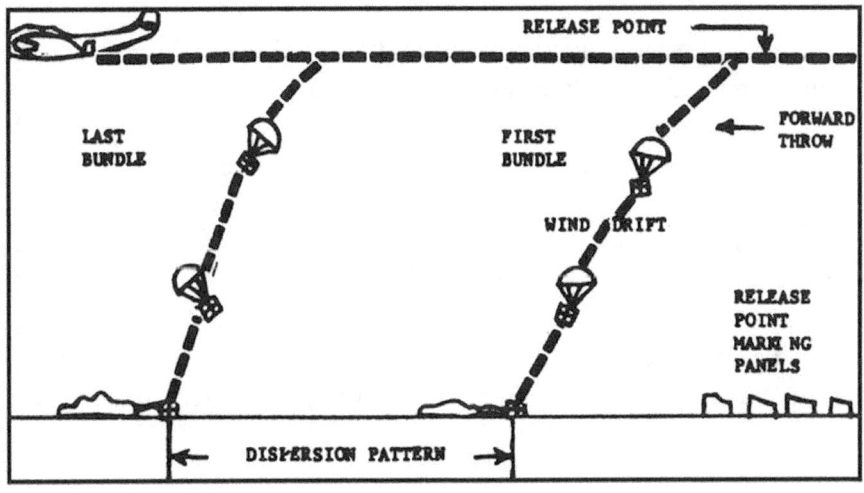

Length of dispersion patterns in meters equals 1/2 aircraft speed (in knots) times exit time (in seconds).

Figure 53. Computation of Dispersion.

Dispersion is computed using the rule-of-thumb formula: 1/2 speed of aircraft (knots) × exit time (seconds) = dispersion (meters). Exit time is the elapsed time between the exits of the first and last items.

The length of the dispersion pattern represents the absolute minimum length required for DZs. If personnel are to be dropped, a safety factor of at least 100 meters is added to each end of the DZ site.

The width of rectangular-shaped DZs should allow for minor errors in computation of wind drift.

The use of DZs measuring less than 300 ×300 meters should be avoided.

Surface.
The surface of the DZ should be reasonably level and free from obstructions such as rocks, trees, fences, etc. Tundra and pastures are types of terrain which are ideal for both personnel and cargo reception.

Personnel DZs located at comparatively high elevations (6,000 feet (1,640 meters) or higher) should, where possible, utilize soft snow or grasslands, due to the increased rate of parachute descent.

Swamps and low marshy ground, normally less desirable in the summer, and paddy fields when dry often make good drop zones.

Personnel and cargo can be received on water DZs.

Minimum depths for reception of personnel is 4 feet and arrangements must be made for rapid pickup.

The surface of the water must be clear of floating debris or moored craft, and there should be no protruding boulders, ledges, or pilings.

The water must also be clear of underwater obstructions to a depth of 4 feet.

Water reception points should not be near shallows or where

currents are swift.

Minimum safe water temperature is 50°F. (10°C.)

Supply drop zones may, in general, utilize any of the following types of surfaces:

Surfaces containing gravel or small stones no larger than a man's fist.

Agricultural ground, although in the interest of security, it is inadvisable to use cultivated fields.

Sites containing brush or even tall trees; however, marking of the DZ and the recovery of containers is more difficult.

Marsh, swamp, or water sites, provided the depth of water or growth of vegetation will not result in loss of containers.

Ground Security.
The basic considerations for ground security are that the DZ be:
Located to permit maximum freedom from enemy interference.

Isolated or in a sparsely populated area.

Accessible to the reception committee by concealed approach and withdrawal routes.

Adjacent to areas suitable for the caching of supplies and disposition of aerial delivery equipment.

III.   REPORTING DROP ZONES:

Drop Zone Data. The minimum drop zone data which is reported includes:

Code name. Extracted from the SOI, also, indicate if primary or alternate DZ.

Location. Complete military grid coordinates of the center of the DZ.

Open Quadrant. Measured from center of DZ, reported as a series of magnetic azimuths. The open quadrant indicates acceptable aircraft approaches (Figure 54).

Track. Magnetic azimuth of required or recommended aircraft approaches (Figure 54).

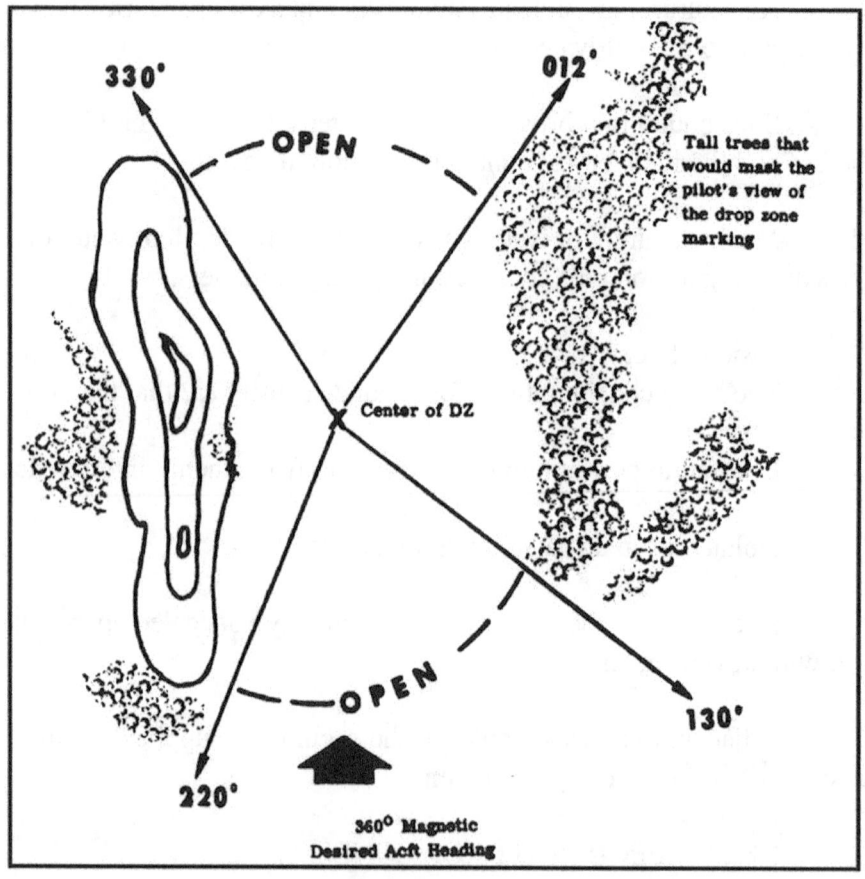

Open Quadrants above would be reported as:
OPEN 130 to 220 AND 330 to 012 DEG
Figure 54. Computation of Open Quadrant

Obstacles. Those that are over 300 feet (90 meters) in elevation above the level of the DZ, within a radius of 6 miles (8 kilometers) and which are not shown on the issued maps. Obstacles are reported by description, magnetic azimuth, and distance from the center of the DZ (Figure 55).

Reference point. A landmark shown on the issued maps, reported by name, magnetic azimuth and distance from the center of the DZ (Figure 55). Used with (2) above in plotting the DZ location.

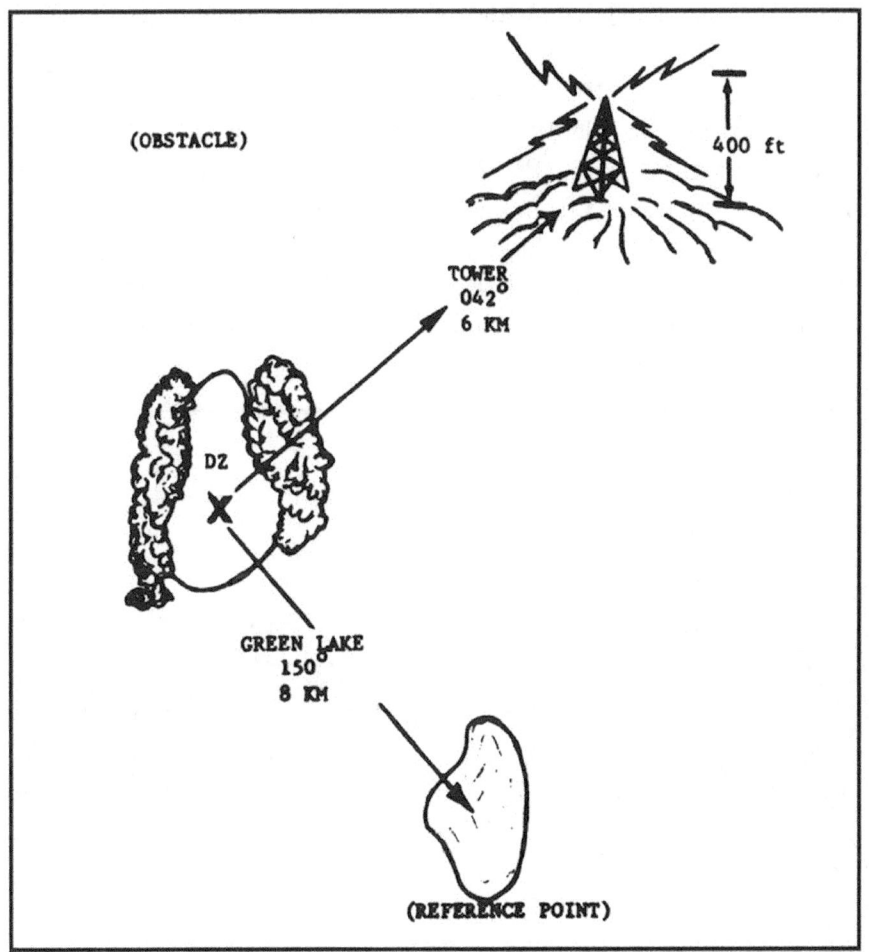

Figure 55. Reporting obstacles and reference points.

Date/time drop requested.

Items requested. Extracted from the catalog supply system.

Additional Items. In special situations, additional items may be required, e.g., additional reference points, navigational check points in the vicinity of the DZ, special recognition and authentication means. Sub-paragraphs (7) and (8) above are included only when requesting a resupply mission in conjunction with the reporting of the DZ.

Azimuths. Azimuths are reported as magnetic and in three digits. With the exception of the aircraft track, all azimuths are measured from the center of the DZ. Appropriate abbreviations are used.

Initial Points (IPs). It is desirable to reconcile the requested aircraft track with an identifiable landmark that may be used by the aircrew as an initial point (IP). The IP, located at a distance of 5 to 15 miles (8-24 kilometers) from the DZ, is the final navigational checkpoint prior to reaching the target. Upon reaching the IP, the pilot turns to a predetermined magnetic heading that takes him over the DZ within a certain number of minutes (Figure 5). The following features constitute suitable IPs:

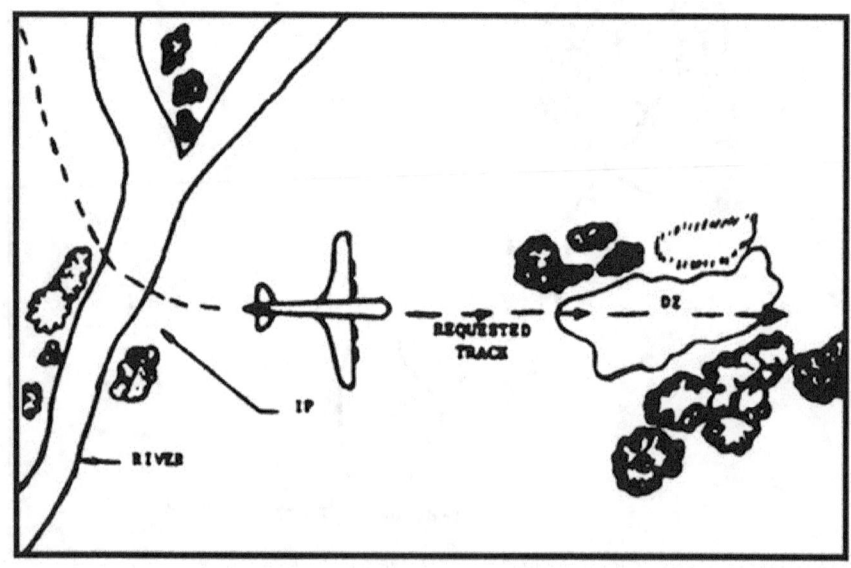

Figure 56. Relationship between IP and requested track.

Coastlines. A coastline with breaking surf is easily distinguished at night. Mouths of rivers over 50 yards wide, sharp uprisings, and inlets are excellent guides for both day and night.

Rivers and canals. Wooded banks reduce reflections, but rivers more than 30 yards wide are visible from the air. Canals are easily recognizable from their straight banks and uniform width. Small streams are not discernible at night.   Figure 5. Relationship between IP and requested track.

Lakes. Lakes at least one-half mile (1 kilometer) square give good light reflection.

Forest and woodlands. Forested areas at least one-half mile square with clearly defined boundaries of unmistakable shape.

Major roads and highways. Straight stretches of main roads with one or more intersections. For night recognition, dark surfaced roads are not desirable as IPs although when the roads are wet, reflection from moonlight is visible.

IV.   MARKING DROP ZONES:

Purpose. The purpose of DZ markings is to identify the site for the aircrew and to indicate the point over which the personnel and/or cargo should be released (release point). The procedures for marking DZs are determined prior to infiltration and are included in the SOI.

Equipment. The marking of DZs at night during clandestine operations will normally be only by flashlights. Flashlights manufactured in the country are easily procured by the guerrillas, give adequate directional lighting when properly held, and are not incriminating when found by the security forces on the person of a member of the resistance force. In rare instances other possible lighting devices such as flares, flarepots, fuses, or small wood fires may be used.

For daylight operations a satisfactory method is the use of issued Panel Marking Set AP-50 or VS-16. If issued panels are not available, sheets, strips of colored cloth or other substitutes may be issued as long as there is a sharp contrast with the background. Smoke signals, either smoke grenade or simple smudge fires, greatly assist the aircrew in sighting the DZ markings on the approach run.

The use of electronic homing devices permits the conduct of

reception operations during conditions of low visibility. Such devices normally are used in conjunction with visual marking systems.

Computation of Release Point. The release point must be determined to insure delivery of personnel and/or cargo within the usable limits of the DZ. Computation of the release point involves the following factors (Figure 57).

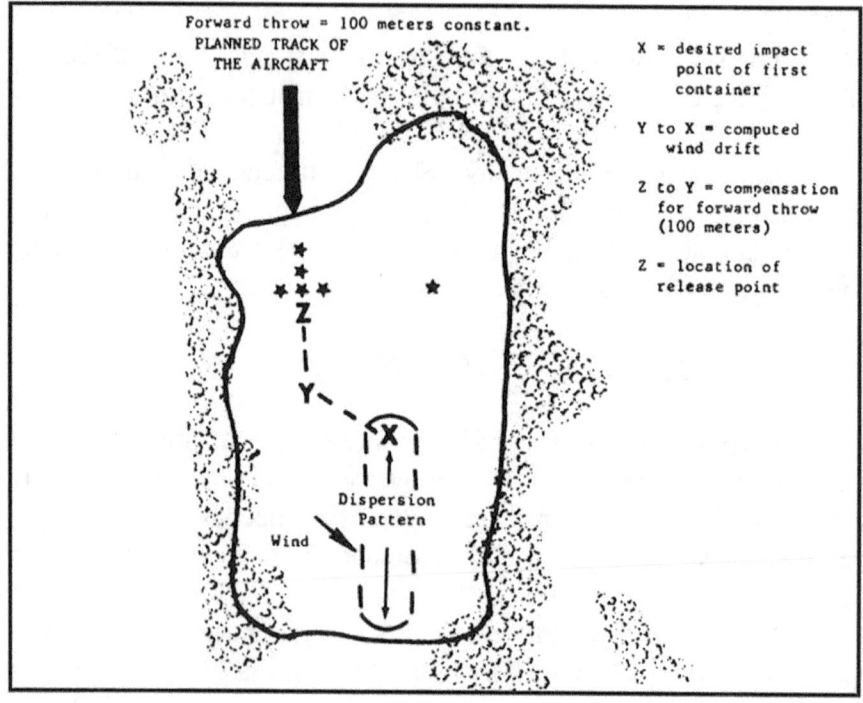

Figure 57.

Personnel from low velocity cargo drops.

Dispersion. Dispersion is the length of the pattern formed by the exit of the parachutists and/or cargo containers (Figure 53). The desired point of impact for the first parachutist/container depends upon the calculated dispersion.

Wind drift. This is the horizontal distance traveled from the point of exit to the point of landing as a result of wind conditions. The release point is located an appropriate distance upwind from the desired impact point. To determine the amount of drift, use the following formulas:

For personnel using the T-10 parachute: Drift (meters) = altitude (hundred of feet) × wind velocity (knots) × 4.1 (constant factor).

For all other low velocity parachute drops: Same as 1 above, however, substitute a constant factor of 2.6 for 4.1.

NOTE: Where no mechanical wind velocity indicator is available, the approximate velocity can be determined by dropping bits of paper, leaves, dry grass, or dust from the shoulder and pointing to the dry place where they land. The estimated angle in degrees formed by the arm with the body, divided by 4, equals wind velocity.

Forward throw.   This is the horizontal distance traveled by the parachutist or cargo container between the point of exit and the opening of the parachute. This factor, combined with reaction time of personnel in the aircraft, is compensated for by moving the release point an additional 100 meters in the direction of the aircraft approach (Figure 57).

High velocity and free-drops. Due to their rapid rate of descent, high velocity and free-drop loads are not materially affected by wind conditions. Otherwise, the factors of dispersion and forward throw are generally similar to those for personnel and low velocity drops and are compensated for in the same manner.

Methods of Release Point Marking. There are two methods for marking the DZ release point. The principal difference between the two is the method of providing identification. The marking systems described below are designed primarily for operational drops executed at an absolute altitude of 600 feet (185 meters). Training jumps executed at an absolute altitude of 1,250 feet (383 meters) require a modification of the marking systems.

Training jumps conducted at an absolute altitude of 1,250 feet (385 meters) require the use of a flank panel or light placed 200 meters to the left of the release point markings. The configuration of present cargo and troop carrying aircraft prevents the pilot from seeing the markings after approaching within approximately one (1) mile of the

DZ while flying at 1,250 feet (385 meters) absolute altitude. From this point on, the pilot must depend on flying the proper track in order to pass over the release point. The flank marker serves to indicate when the aircraft is over the release point and the exact moment the drop should be executed. Operational drops executed at 600 feet (185 meters) absolute altitude do not require the flank panel because the pilot does not lose sight of the markings as he approaches the DZ. (See Figure 58.)

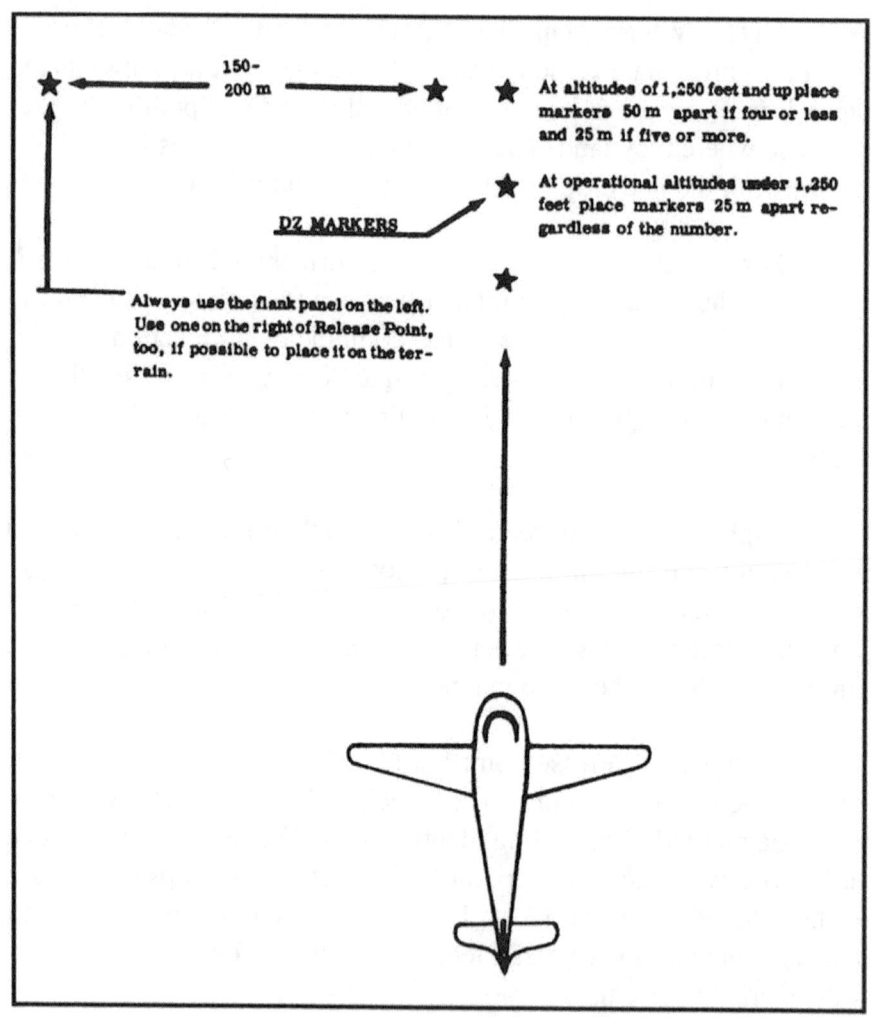

Figure 58. Methods of Release Point Marking.

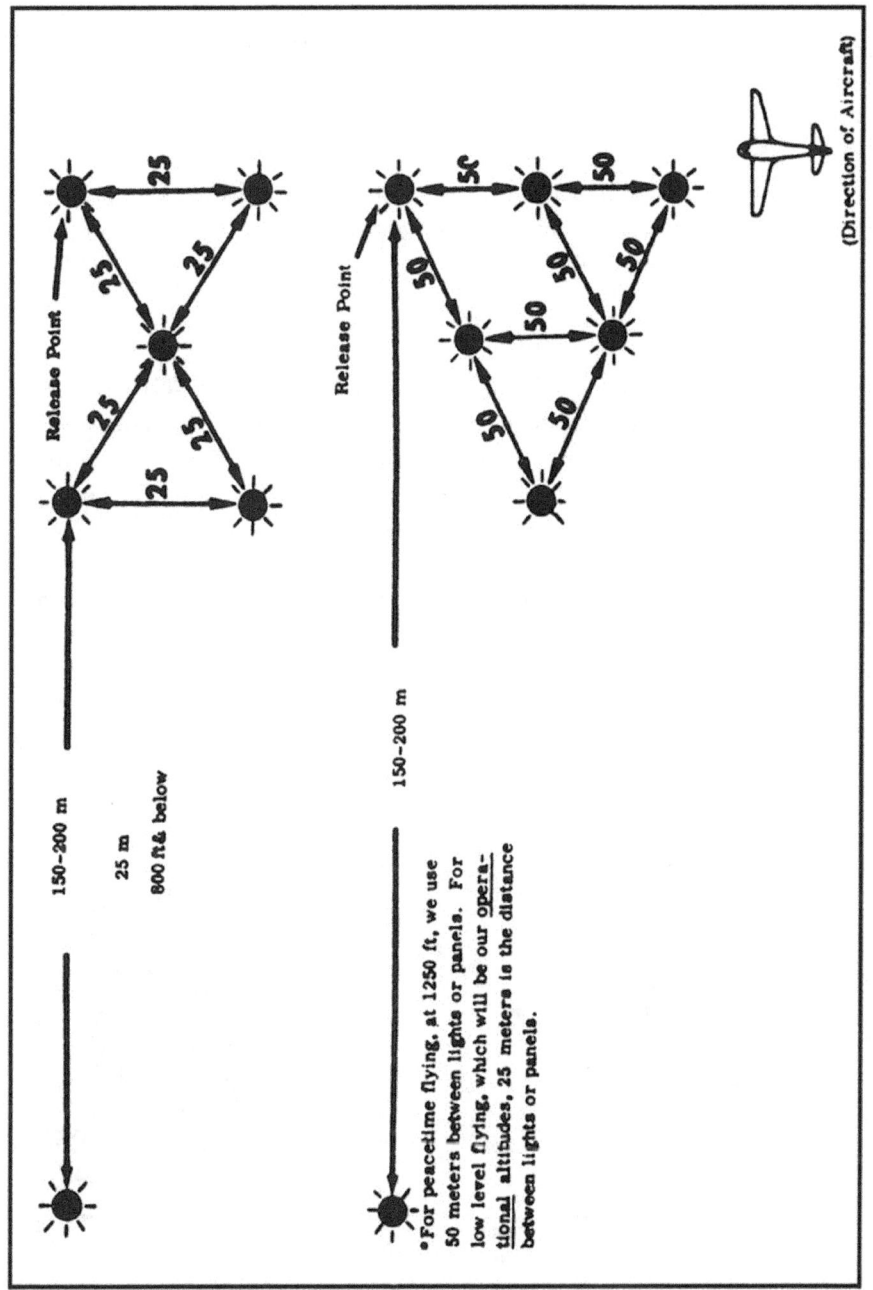

Figure 59. Sample DZ Markings.
(Distinctive configuration, not necessarily a letter.)

Operational personnel drops or supply drops within a GWOA will normally be executed at altitudes between 600-800 feet for personnel and 400-600 feet for supplies. Release point markings are different numbers of lights with different configurations for each 24-hour period. The exact number of lights and the exact configuration is determined by the detachment SOI. (See Figure 59.)

Placement of Markings. Markings must be clearly visible to the pilot of the approaching aircraft. As a guide, markings must have a clearance of at least 500 yards (460 meters) from a 100-foot (30 meter) mark (Figure 60).

Additionally, precautions must be taken to insure that the markings can be seen only from the direction of the aircraft approach. Flashlights may be equipped with simple hoods or shields and aimed toward the flight path. Fires or improvised flares are screened on three sides or placed in pits with sides sloping toward the direction of aircraft approach.

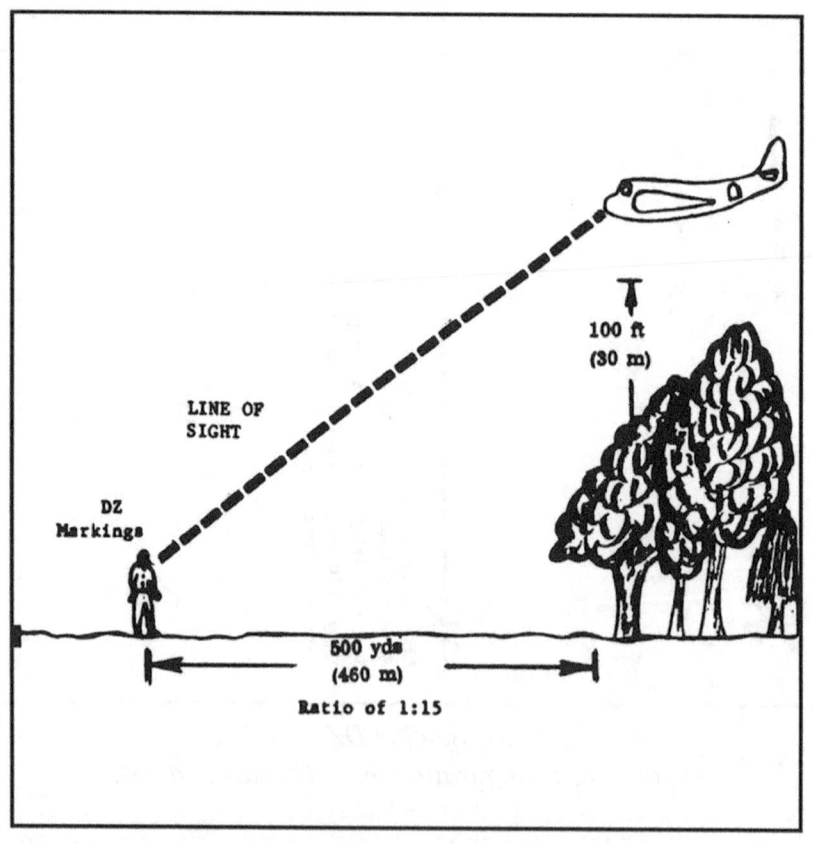

Figure 60. Placement of DZ Markings.

When panels are used for daylight markings of DZs, they are positioned at an angle of approximately 45° from the horizontal to present the maximum surface toward the approaching aircraft (Figure 61).

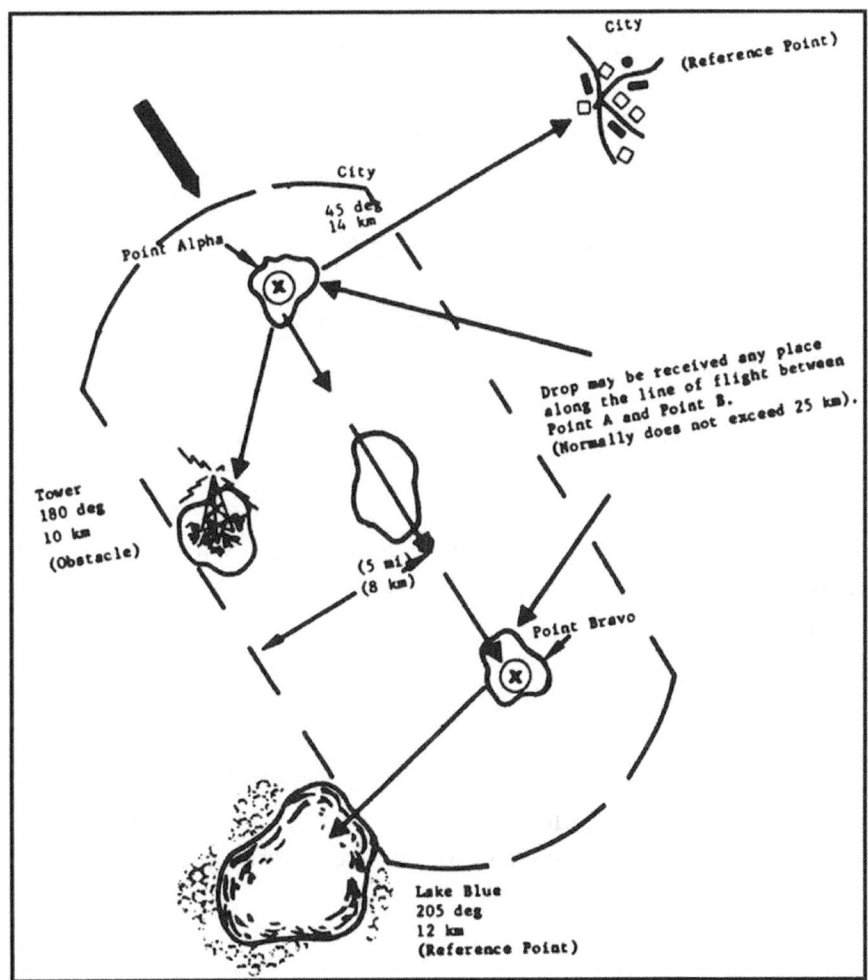

Figure 61. Obstacles and Reference Point (Area DZ).

V.   RECEPTION COMMITTEES:

a. General. A reception committee is formed to control the drop zone or landing area. The reception committee can be anyone who is capable of performing the following duties. A permanent committee for each unit provides the best results, eliminating the need to cross train every one to be capable of this mission. However, training in depth should be accomplished to insure that losses of key personnel will not adversely affect the operation of the group as a whole.

Provide security for the reception operation.

Emplace DZ markings and air ground identification equipment.

Maintain surveillance of the site prior to and following the reception operation.

Recover and dispose of incoming personnel and/or cargo.

Provide for dispatch of personnel and/or cargo in evacuation operations.

Provide for sterilization of the site (when secrecy is possible and desirable only).

c. Composition. The reception committee is normally organized into five parties. The composition and functions of the five parties are as follows:

Command party. Controls and coordinates the actions of all reception committee components.
Includes the reception committee leader (RCL) and communications personnel, consisting of messengers and radio operators.
Provides medical support, to include litter bearers, during personnel drops.

Marking party.
Operates the reception site marking system, using one man for each marker.
The marking party must be well rehearsed. Improperly placed or improperly operated markings may cause an aborting of the mission.

Security party.
Insures that unfriendly elements do not interfere with the conduct of the operation.
Consists normally of inner and outer security elements.

The inner security element is positioned in the immediate vicinity of the site and is prepared to fight delaying or holding actions.

The outer security element consists of outposts established along approaches to the area. They may prepare ambushes and road blocks to prevent enemy movement toward the site.

The security party may be supplemented by auxiliaries. These are generally used to maintain surveillance of enemy activities and keep the security party informed of hostile movements.

Provides march security for moves between the reception site and the destination of the cargo or infiltrated personnel.

Recovery party.

Recovers cargo and aerial delivery equipment from the DZ. Unloads aircraft or landing craft.

For aerial delivery operations the recovery party should consist of at least one man for each parachutist or cargo container. For such operations, the recovery party is usually dispersed along the length of the anticipated Impact area. The members spot each parachute as it descends and move to the landing point. They then recover all parachute equipment and cargo, moving to a predetermined assembly area with the infiltrated personnel or equipment.

The recovery party is normally responsible for sterilizing the reception site to insure that all traces of the operation are removed when secrecy is possible and desired.

Transport party.

Moves items received to distribution points or caches.

May consist of part, or all, of the members comprising the command, marking, and recovery parties.

Uses available means of transportation such as pack animals and wagons.

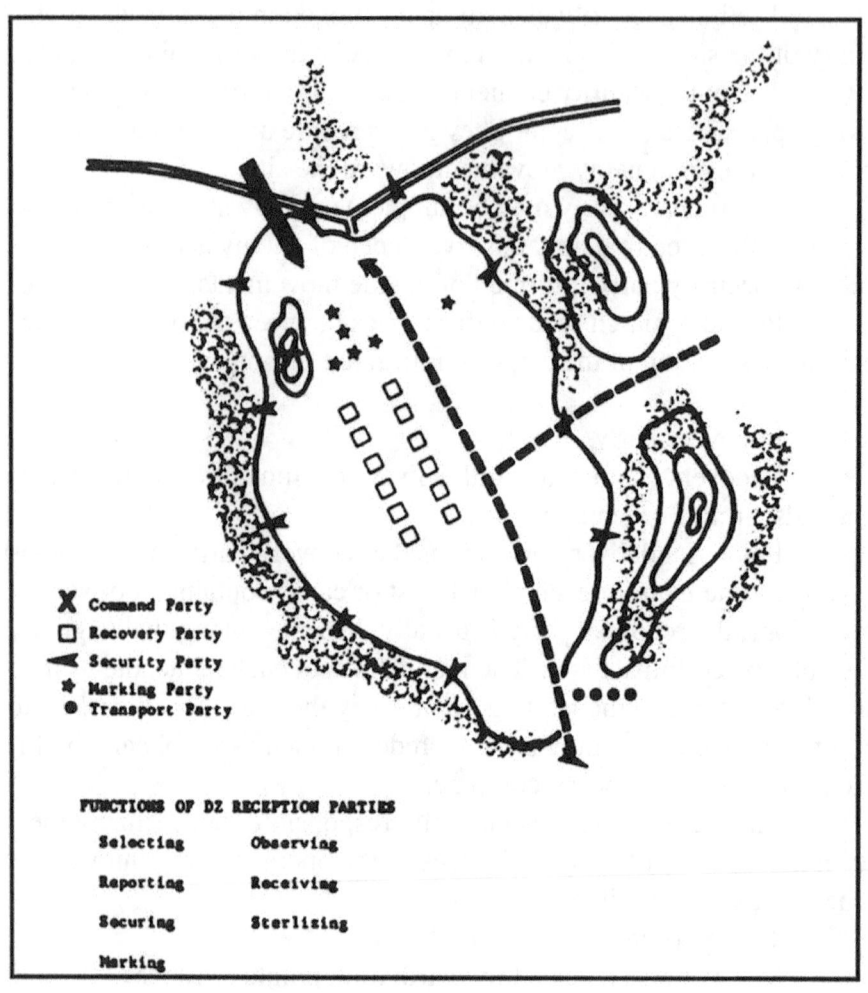

Figure 62. Organization of DZ Reception.

VI.   LANDING ZONES (LAND)

General. The same general considerations applicable to DZ selections apply to the selection of LZs. However, site size, approach features and security are far more important.

Selection Criteria.
Desirable terrain features:
LZs should be located in flat or rolling terrain.
Level plateaus of sufficient size can be used. Due to decreased air density, landings at higher elevations require Increased minimum

LZ dimensions. If the LZ is located in terrain above 4,000 feet (1,220 meters) and/or areas with a very high temperature the minimum lengths should be increased as follows:

Add 10 percent to minimums for each 1,000 feet (305 meters).

Add 10 percent to minimum for the altitude for temperatures over 90°F.

Add 20 percent for temperatures over 100°F. (38°C).

Pockets or small valleys completely surrounded by hills are usually unsuitable for landing operations by fixed-wing aircraft.

Although undesirable, sites with only a single approach can be used. It is mandatory when using such sites that:

All takeoffs and landings are made upwind.

There is sufficient clearance at either end of the LZ to permit a level 180° turn to either side within a radius of 3 miles (5 kilometers) for medium aircraft (1 mile for light aircraft).

Weather. Prevailing weather in the landing area should be favorable. In particular, there must be a determination of wind direction and velocity, and of conditions restricting visibility such as ground fog, haze, or low-hanging cloud formations.

Size. The required size of LZs varies according to the aircraft used. Safe operations require the following minimum dimensions (Figures 63 and 64).

Medium aircraft. 3,000 feet (920 meters) in length and 100 feet (30 meters) in width (150 feet or 45 meters at night).

Light aircraft. 1,000 feet (305 meters) in length and 40 feet (15 meters) in width (150 feet or 45 meters at night).

In addition to the basic runway dimensions, and to provide a safety factor, these extra clearances are required.

A cleared surface capable of supporting the aircraft, extending from each end of the runway, and equal to 10 percent of the runway length.

A 50-foot (15 meter) strip extending along both sides of the runway and cleared to within three feet of the ground.

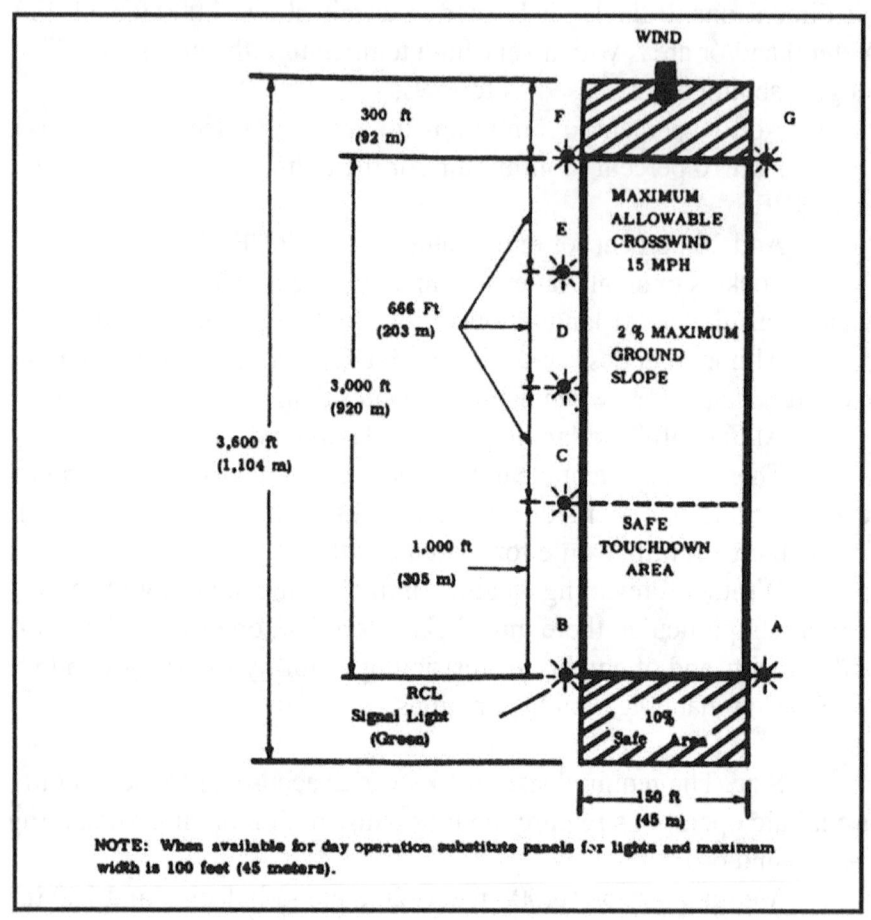

WIND

300 ft (92 m)

F   G

MAXIMUM ALLOWABLE CROSSWIND 15 MPH

E

666 Ft (203 m)

D   2 % MAXIMUM GROUND SLOPE

3,000 ft (920 m)

3,600 ft (1,104 m)

C

SAFE TOUCHDOWN AREA

1,000 ft (305 m)

B   A

RCL Signal Light (Green)

10% Safe Area

150 ft (45 m)

NOTE: When available for day operation substitute panels for lights and maximum width is 100 feet (45 meters).

Figure 63. Landing zone (land) medium aircraft (night operations).

Surface.

The surface of the LZ must be level and free of obstructions such as ditches, deep ruts, logs, fences, hedges, low shrubbery, rocks larger than a man's fist or grass over 1 1/2 feet in height.

The sub-soil must be firm to a depth of 2 feet.

A surface-containing gravel and small stones, or thin layers of loose sand over a firm layer of sub-soil is acceptable. Plowed fields or fields containing crops over 1 1/2 feet in height should not be used.

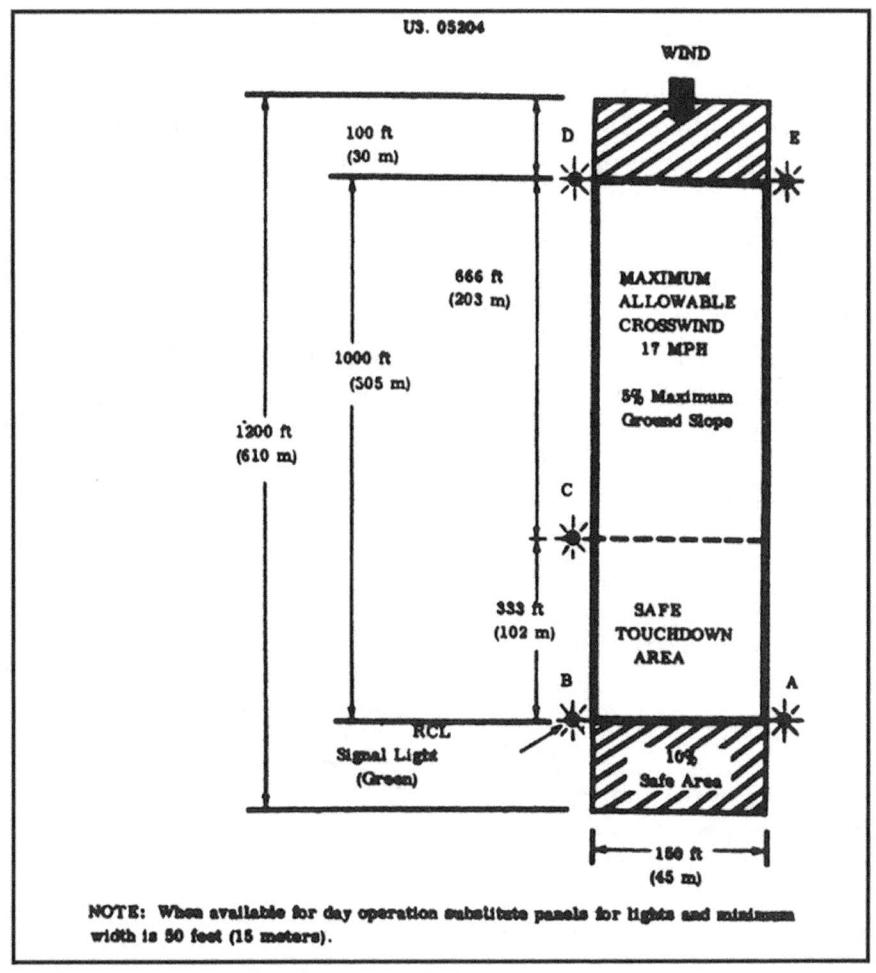

Figure 64. Landing zone (land) light aircraft (night operations).

As with DZs surfaces that are not desirable in summer may be ideal In winter. Ice with a thickness of 2 feet (61 centimeters) will support a medium aircraft. Unless the aircraft is equipped for snow landing, snow in excess of 4 inches (11 centimeters) must be packed or removed from the landing strip.

The surface gradient of the LZ should not exceed 2 percent.

Approach and takeoff clearance. The approach and takeoff clearances are based on the glide-climb characteristics of the aircraft. For medium aircraft the glide-climb ratio is 1 to 40; that is, 1 foot of gain or loss of altitude for every 40 feet of horizontal distance traveled. The ratio for light aircraft is 1 to 20. As a further precaution, any obstructions in approach and departure lanes must conform to the follow-

85

ing specifications (Figure 65).

An obstruction higher than 6 feet (2 meters) is not permissible at or near either end of the LZ.

A 50-foot (15 meters) obstruction may not be nearer than 2,000 feet (610 meters) for medium aircraft, or 1,000 feet (305 meters) for light aircraft.

A 500-foot (155 meter) obstruction may not be nearer than 4 miles (617 kilometers) for medium aircraft or 2 miles (305 meters) for light aircraft.

Hills of 1,000 (305 meters) feet or more above LZ altitude may not be nearer than 8 miles (13 kilometers) from the landing sons for medium aircraft.

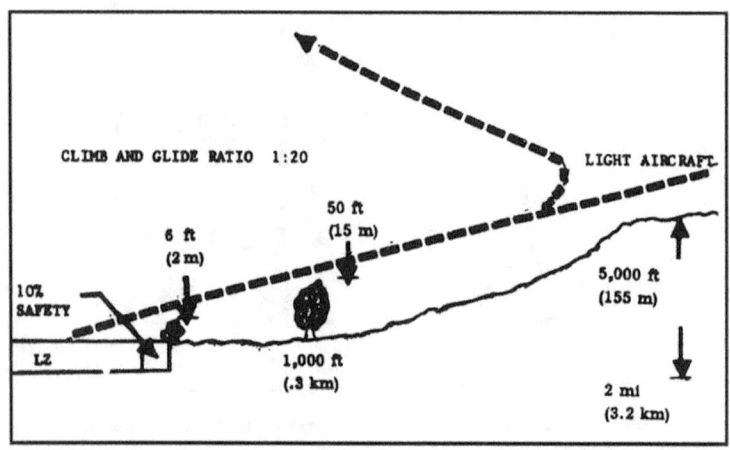

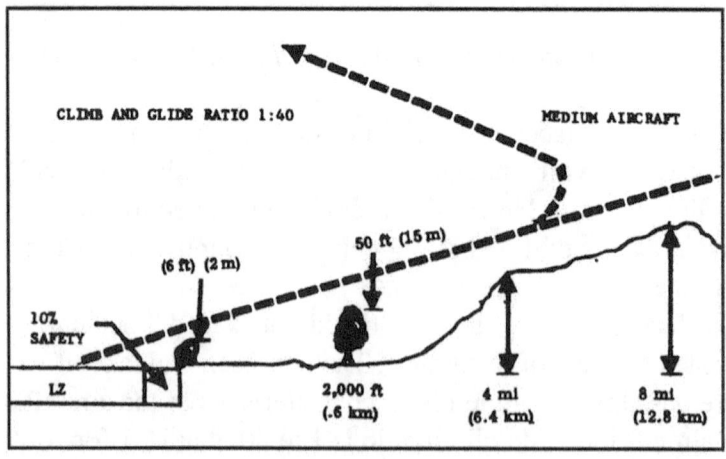

Figure 65.

The heights of the obstacles are computed from the level of the landing strip. Where land falls away from the LZ, objects of considerable height may be ignored provided they do not cut the line of ascent or descent. This condition exists more often in mountainous terrain where plateaus are selected for LZs.

Markings.

For night operations lights are used for marking LZs; during day-light, panels are used. When flashlights are used, they should be hand-held for directional control and guidance.

The pattern outlining the limits of the runway consists of five or seven marker stations (Figures 11 and 12). Stations "A" and "B" mark the downwind end of the LZ and are positioned to provide for the safety factors previously mentioned. These stations represent the initial point at which the aircraft should touch the ground. Station "C" indicates the very last point at which the aircraft can touch down and complete a safe landing.

A signal station manned by the RCL (a member of the operational detachment) is incorporated into light station "B" at the approach on downwind end of the LZ (Figures 11 and 12). For night operations, (the signal light operations,) a distinctive panel or colored smoke, located approximately 15 meters to the left of station "B" (RCL), is used for recognition.

Conduct of Operations.

The LZ markings are normally displayed 2 minutes before the arrival time indicated in the mission confirmation message. The markings remain displayed for a period of 4 minutes or until the aircraft completes landing roll after touchdown.

Identification is accomplished by:

The aircraft arriving at the proper time on prearranged track.

The reception committee leader flashing or displaying the proper code signal.

Landing direction is indicated by:

The RCL signal control light (station "B") and marker "A" which are always on the approach or downwind end of the runway.

The row of markers which are always on the left side of the landing aircraft.

The pilot usually attempts to land straight-in on the initial approach. When this is not possible, a modified landing pattern is flown using a minimum of altitude for security reasons. Two minutes before target time the RCL causes all lights of the LZ pattern to be turned on and aimed like a pistol in the direction of the aircraft's approach track. The RCL (station "B") also flashes the code of the day continuously with the green control light in the direction of expected aircraft approach. Upon arrival in the area (within 15° to either side of the approach track and below 1,500 feet (460 meters)), the LZ marking personnel follow the aircraft with all lights when it arrives in the area. When the RCL determines that the aircraft is on its final approach, he will cease flashing the code of the day and aim a solid light in the direction of the landing aircraft. The solid light provides a more positive pattern perspective for the pilot during landing. If a "go around" is required, all lights follow the aircraft until it is on the ground. All lights continue to follow the aircraft during touchdown and until it passes each respective light station.

Landings are not normally made under the following conditions:

Lack of or improper identification received from the LZ.

An abort signal given by the RCL, e.g., causing the LZ lights to be extinguished.

Any existing condition that, in the opinion of the pilot, makes it unsafe to land.

After the aircraft passes the RCL position at touchdown and completes its landing roll and a right turn, the RCL takes a position midway between stations "A" and "B" and shines a solid light in the direction of the taxiing aircraft. This is the guide light for the pilot who will taxi the aircraft back to take-off position. The RCL controls the aircraft with his light. If the RCL desires the aircraft to continue to taxi, he will flash a solid light in the direction of the aircraft. After off-loading and/or on-loading is complete and the aircraft is ready for takeoff, the RCL moves to a vantage point forward and to the left of the pilot, causes the LZ lights to be illuminated, and flashes his light toward the nose of the aircraft as the signal for takeoff. The RCL exercises caution so that his light does not blind the pilot.

To eliminate confusion and insure expeditious handling, personnel and/or cargo to be evacuated wait for unloading of incoming personnel and/or cargo.

When all evacuating personnel are loaded and members of the

reception committee are clear of the aircraft, the pilot is given a go signal by the RCL. LZ markings are removed as soon as the aircraft is airborne.

VII.   REPORTING LANDING ZONES.

The minimum LZ data required is:

Code Name. Extracted from SOI.

Location. Complete military grid coordinates of center of LZ.

Long Axis. Magnetic azimuth of long axis of runway. It also indicates probable direction of landing approach based on prevailing winds.

Description. Type of surface, length, and width of runway.

Open Quadrant. Measured from center of LZ and reported as series of magnetic azimuths. Open Quadrant indicates acceptable aircraft approaches.

Track. Magnetic azimuth of desired aircraft approach.

Obstacles. Reported by description, magnetic azimuth, and distance from center of LZ.

Reference Point. Reported same as obstacles.

Date. Time mission requested.

Items Requested. Items to be evacuated.

VIII.   LANDING ZONES FOR ROTARY-WING AIRCRAFT:

General.

Within their range limitations, helicopters provide an excellent means of evacuation. Their advantages include the ability to:

Ascend and descend almost vertically.

Land on relatively small plots of ground.

Hover nearly motionless, and take on or discharge personnel and cargo without landing.

Fly safely and efficiently at low altitudes.

Some unfavorable characteristics of helicopters are:

They compromise secrecy by engine and rotor noise and by dust.

The difficulty—sometimes impossibility—of operating when icing and/or high, gusty winds prevail.

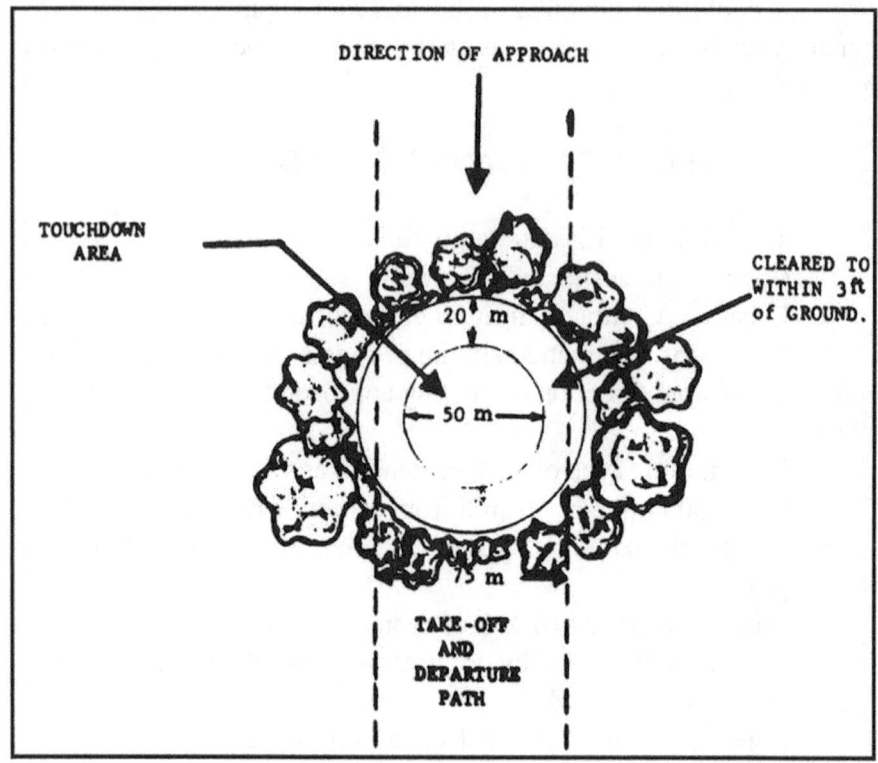

Figure 66. Landing Zone for Rotary-Wing Aircraft.

The reduction of lifting ability during changes of atmospheric conditions.

For the maximum effective use of helicopters, LZs should be located to have landings and takeoffs into the wind.

During night operations, helicopters usually must land to transfer personnel and/or cargo.

A decrease in normal air density limits the helicopter payload and requires lengthened running distances for landing and takeoff. Air density is largely determined by altitude and temperature. Low altitudes and moderate to low temperatures result in increased air density.

Size. Under ideal conditions, and provided the necessary clearance for the rotors exists, a helicopter can land on a plot of ground slightly larger than the spread of its landing gear. For night operations, however, a safety factor is allowed with the following criteria as a guide:

An area of 50 meters in diameter cleared to the ground.

An area beyond this, surrounding the cleared area, 20 meters wide and cleared to within 3 feet of the ground.

The completed LZ is thus a minimum of 90 meters in diameter (Figure 66).

Surface.

The surface should be relatively level and free of obstructions such as rocks, logs, tall grass, ditches, and fences.

The maximum ground slope permitted is 15 percent.

The ground must be firm enough to support the aircraft. Figure 68.

Heavy dust or loose snow conditions interfere with the vision of the pilot Just before touchdown. This effect can be reduced by clearing, wetting down, or using improvised mats.

Landing pads may be prepared on swamp or marsh areas by building platforms of locally available materials (Figure 67). Such LZs are normally used for daylight operations only. The size of the clearing for this type of LZ is the same as above, with the following additional requirements for the platform:

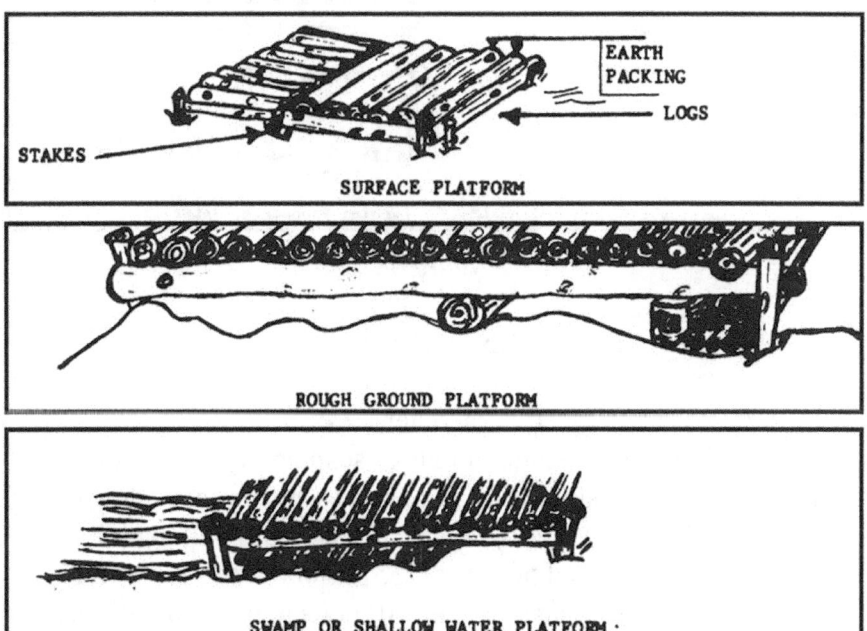

Figure 67. Example of platform landing zones for rotary-wing aircraft.

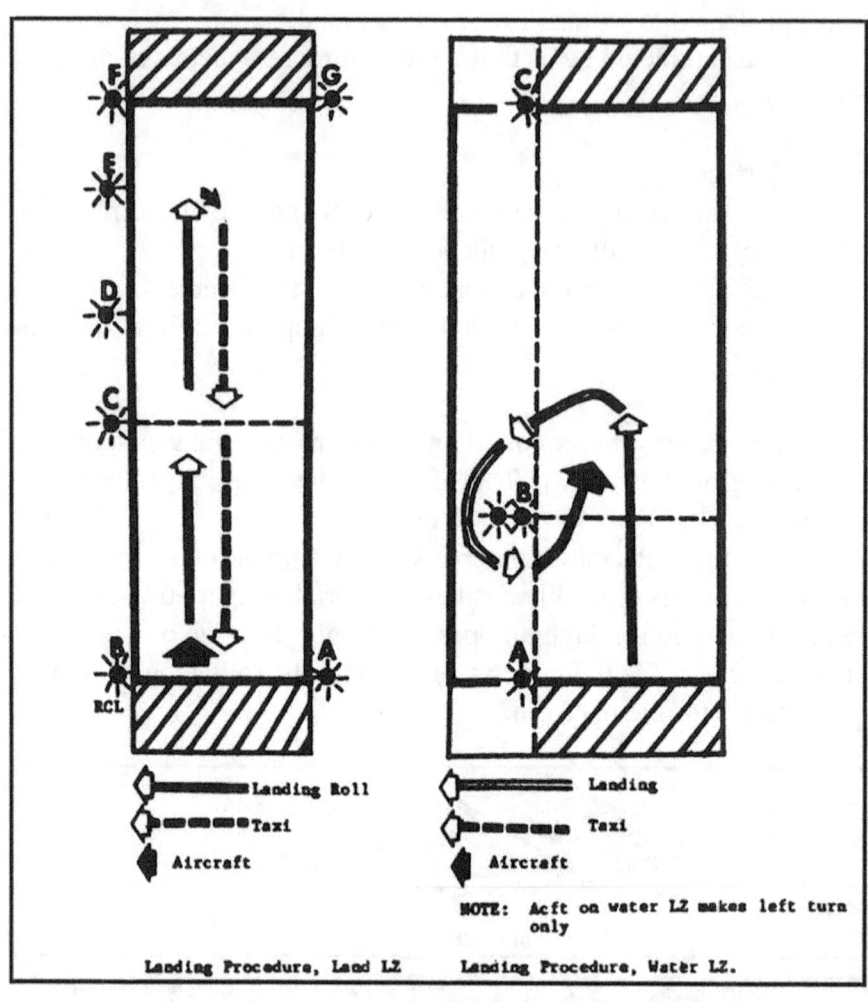

Figure 68.

IX. LANDING ZONES (WATER)

a. Criteria for selection of water LZs:

Size. For medium amphibious or seaplane-type aircraft, the required length is 4,000 feet (1,220 meters) with a minimum width of 1,500 feet (460 meters). For light aircraft, the required length is 2,000 feet (615 meters) long and 500 feet (155 meters) wide. As with land LZs, and additional safe area equal to 10 percent of the airstrip length is required on each end. (Figure 69.)

Surface. Minimum water depth is 6 feet (2 meters). The entire landing zone must be free of obstructions such as boulders, rock

92

ledges, shoals, waterlogged boats, or sunken pilings within 6 feet of the surface, and the surface must be cleared of all floating objects such as logs, debris, or moored craft.

Wind. Wind velocity must not exceed 20 knots for sheltered water or 10 knots in semi-sheltered water.

In a wind of 8 knots or less, the landing heading may vary up to 15 degrees from the wind direction. Where the surface winds exceed 8 knots the aircraft must land into the wind. No landing may be made in winds in excess of 20 knots. If a downwind landing or takeoff is absolutely required, this is made directly downwind.

Surface swells must not exceed 1 foot in height and the wind-wave not more than 3 feet. The combination of swell and windwave must not exceed 3 feet in height when all swells and windwaves are in phase.

Tide. The state of the tide should have no bearing on the suitability of the landing area. Water/air temperature. Due to the danger of icing, water and air temperatures must conform to the following minimums:

	Water temperature	Air temperature
Salt water	−18°F. (−8°C.)	−26°F. (−3°C.)
Fresh water	−35°F. (−2°C.)	−35°F. (−2°C.)
Brackish water	−30°F. (−1°C.)	−35°F. (−2°C.)

Approach and takeoff clearances. Water landing zones require approach/takeoff clearances identical to those of land LZs and are based on the same glide/climb ratios.

c. Marking and identification of water landing zones.

Depending upon visibility, lights or panels may be used to mark water LZs.

The normal method of marking water LZs is to align three marker stations along the left edge of the landing strip. Station "A" is positioned at the downwind end of the strip and indicates the desired touchdown point. Station "B" marks the last point at which the aircraft can touch down and complete a safe landing. Station "B" is also the location of the RCL and the pickup point. Station "C" marks the upwind extreme of the landing area. At night, stations "A," "B," and "C" are marked by white lights. The RCL signal light is green.

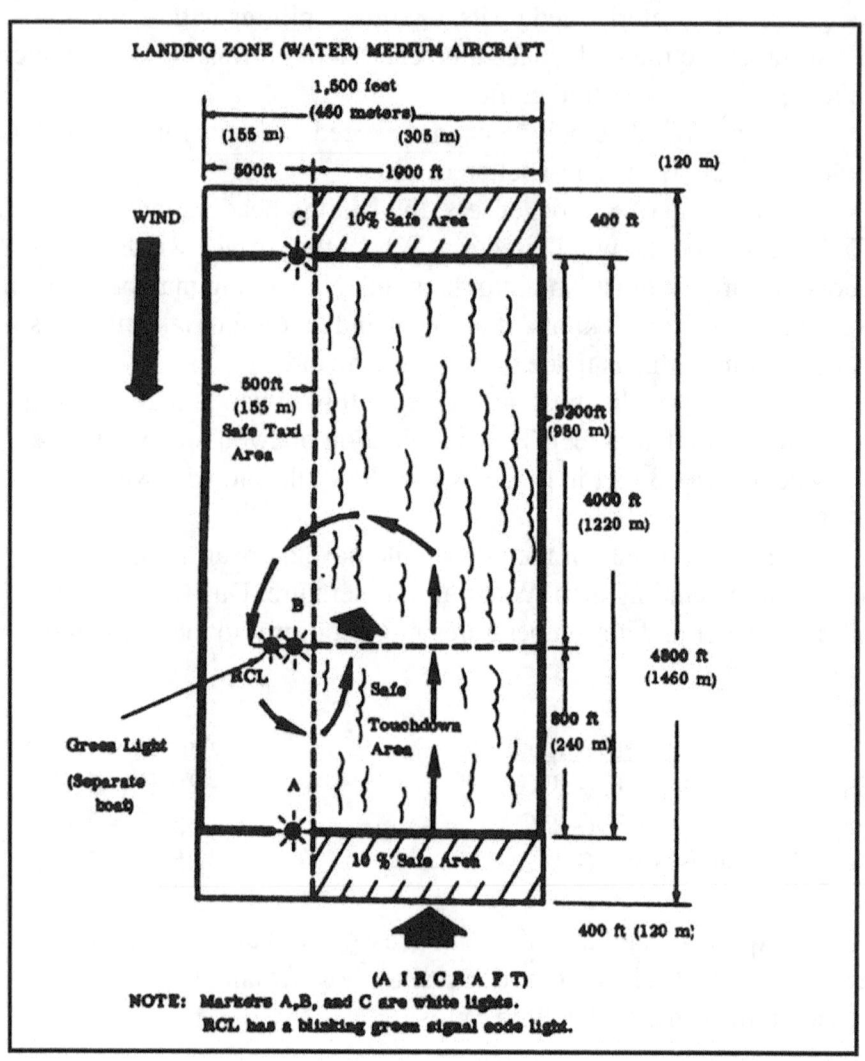

Figure 69. Landing Zone (water) medium aircraft (night operations).

An alternate method is to use a single marker station, marked at night with a steady light in addition to the signal of recognition light. This station is located to allow a clear approach and takeoff in any direction. The pilot is responsible for selecting the landing track and may touchdown on any track 1,000 feet (305 meters) from the marker station. Following pickup, the aircraft taxis back to the 2,000-foot (610 meters) circle in preparation for takeoff. (Figure 70.)

d. Conduct of operations for water LZs:

94

Before the landing operation, the LZ is carefully cleared of all floating debris. Also, the marker stations are properly aligned and anchored to prevent drifting. In deep or rough water, improvised sea anchors may be used.

The procedure for displaying the LZ markings and identification is the same as for operations on land LZs.

Personnel and/or cargo to be evacuated are positioned in the RCL boat. Following the landing run, the aircraft turns to the left and taxis back to the vicinity of the RCL boat to make the pickup. The RCL indicates his position by shining the signal light in the direction of the aircraft and continues to shine his light until the pickup is completed. Care must be taken not to blind the aircrew with this light and it should not be aimed directly into the cockpit.

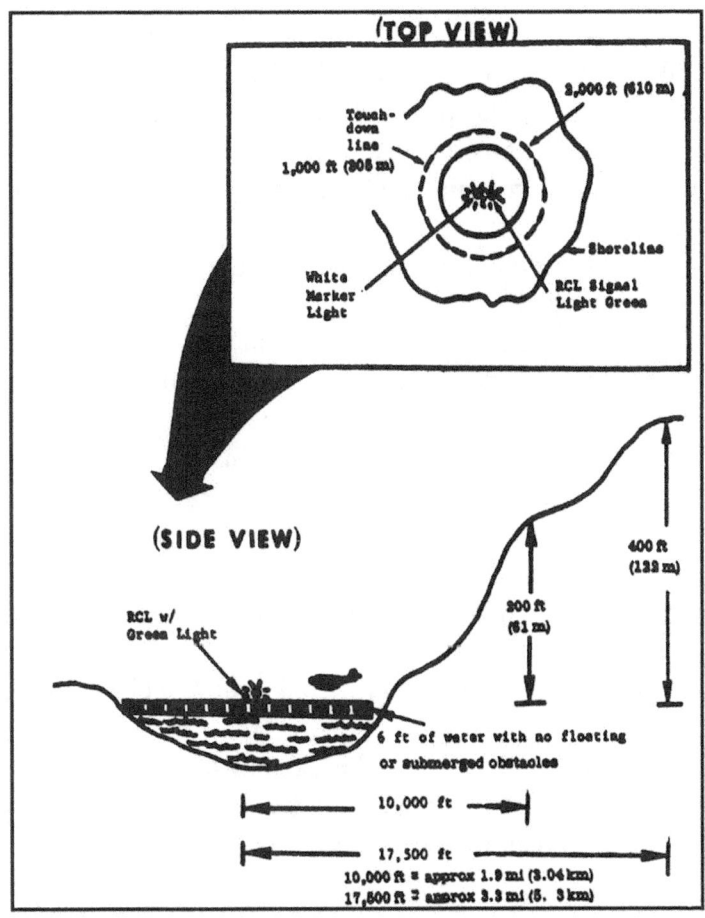

Figure 70. One light water landing zone (night).

The RCL boat remains stationary during pickup operations. The aircraft taxis to within 50 to 100 feet (15 or 30 meters) of the RCL boat, playing out a dragline from the left rear door. The dragline is approximately 150 feet (45 meters) in length and has three life jackets attached; one close to the aircraft, a second at midpoint, and the third on the extreme end of the line. The life jackets have small marker lights attached during night operations. The aircraft taxis to the left around the RCL boat, bringing the dragline close enough to be secured. The RCL fastens the line to the boat. Due to the danger of swamping the craft, the RCL does not attempt to pull on the line. Members of the aircrew pull the boat to the door of the aircraft. Should the boat pass the aircraft door and continue toward the front of the aircraft, all personnel in the boat must abandon immediately to avoid being hit by the propeller.

After pickup, the aircrew is given any information that will aid in the takeoff. Following this, the RCL boat moves a safe distance from the aircraft and signals the pilot "all clear." At this time, JATO bottles may be used for positive takeoff power. The installation of JATO bottles is time consuming and should not be done unless absolutely necessary.

Helicopters can land in water without the use of special flotation equipment provided:

The water depth does not exceed 18 inches.

There is a firm bottom such as gravel or sand.

Landing pads can be prepared on mountains or hillsides by cutting and filling. Caution must be exercised to insure there is adequate clearance for the rotors.

d. Approach/Takeoff.

There should be at least one path of approach to the LZ measuring 75 meters in width.

A rotary wing aircraft is considered to have a climb ratio of 1:5 (Figure 71).

Takeoff and departure from the LZ may be along the same path used for the approach; however, a separate departure path as free from obstacles as the approach path is desired (Figure 71).

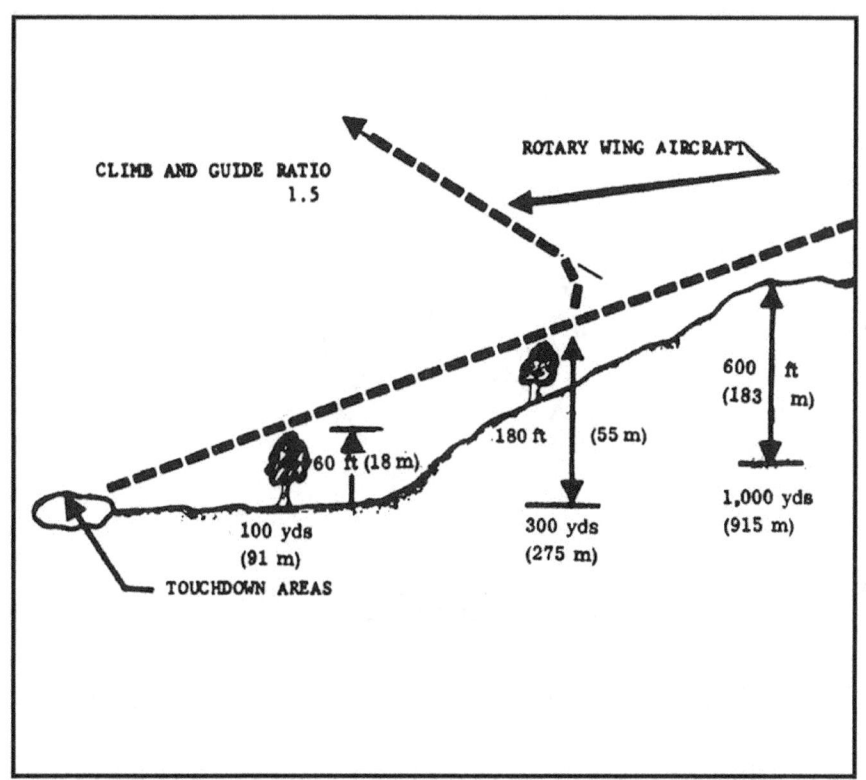

Figure 71. Approaching Takeoff Clearances for Rotary-Wing Aircraft.

e. Marking.

LZs for rotary-wing aircraft are marked to:

Provide identification of the reception committee.

Indicate direction of wind and/or required direction of approach.

Delineate the touchdown area.

Equipment and techniques of marking are similar to those used with fixed-wing DZs, lights or flares at night, smoke and panels in daylight.

An acceptable method of marking is the "Y" system. This uses four marker stations (Figure 72).

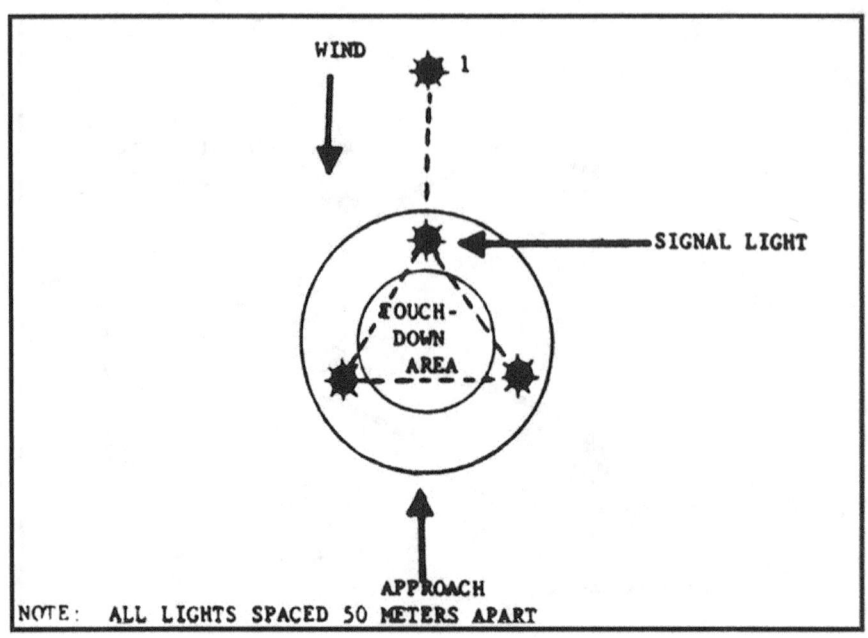

Figure 72. Marking of landing zones for use by rotary-wing aircraft.

TABLE NR. 1. FIXED AND ROTARY WING AIRCRAFT CAPABILITIES.

Air Craft	Cruise Speed (Knots)	Range Full Fuel NM	Payload 50 NM Radius	Payload 100 NM Radius	Trps Cmbr Equipped	Litters	Cu. Ft. Cargo Space	External Sling Capacity
UH-1B	80	210	3000	2600	7	3	140	3000
UH-1D	100	293	3300	2700	11	6	220	4000
CH-21	80	240	4000	3600	20	12	422	5000
CH-34	80	200	4000	3500	18	8	363	5000
CH-37	80	120	6500	4100	23	24	1142	10000
01-	87	390	500	500	1	--	--	--
U-6	105	575	1300	1150	4	2	125	--
U-8F	165	1100	2200	2100	4	4	192	--
U-1A	100	700	2400	2200	9	6	286	--
OV-1	200	410	2000	--	--	--	--	--
CV-2	157	1050	7500	7500	32	14	1150	--

NOTE: Above figures are to be used as guides only. Many factors will influence the capabilities of any aircraft. Increases in temperature, humidity and altitude will decrease performance. Desired range will affect fuel load which will determine number of troops or amount of cargo that can be carried.

CHAPTER 5
WEAPONS

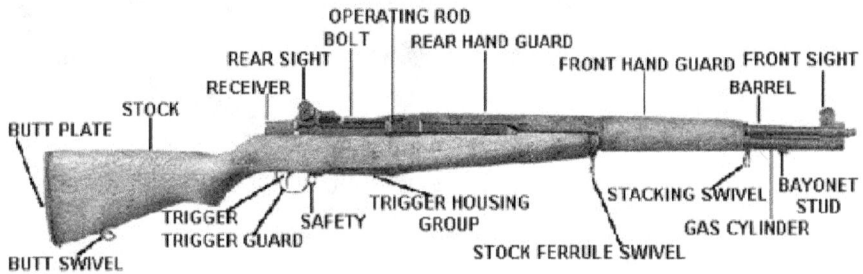

Figure 73. U.S. Rifle Caliber .30 M-1.

Characteristics: Data:

Air cooled   Max. effective range (500yds)
Semi-automatic   Max. range (3450yds)
Gas operated   Clip capacity (8rnds)
Shoulder weapon
Clip loaded

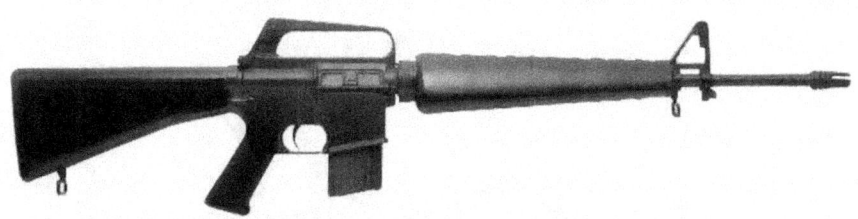

Figure 74. Colt AR-15 Cal .223 (Redesignated M-16 Rifle).

Characteristics: Data:

Gas operated   Max. range (2833 yds)
Air cooled   Max. effective range (500 yds)
Semi or fully automatic
Shoulder weapon
Magazine fed

Figure 75. Carbine Cal .30 M1 & M2.

Characteristics: Data:
Air cooled   Mag. capacity (15/30 rds)
Magazine loaded   Max. effective range (275 yds)
Gas operated   Max. range (2,200 yds)
Semi and fully automatic
Shoulder weapon

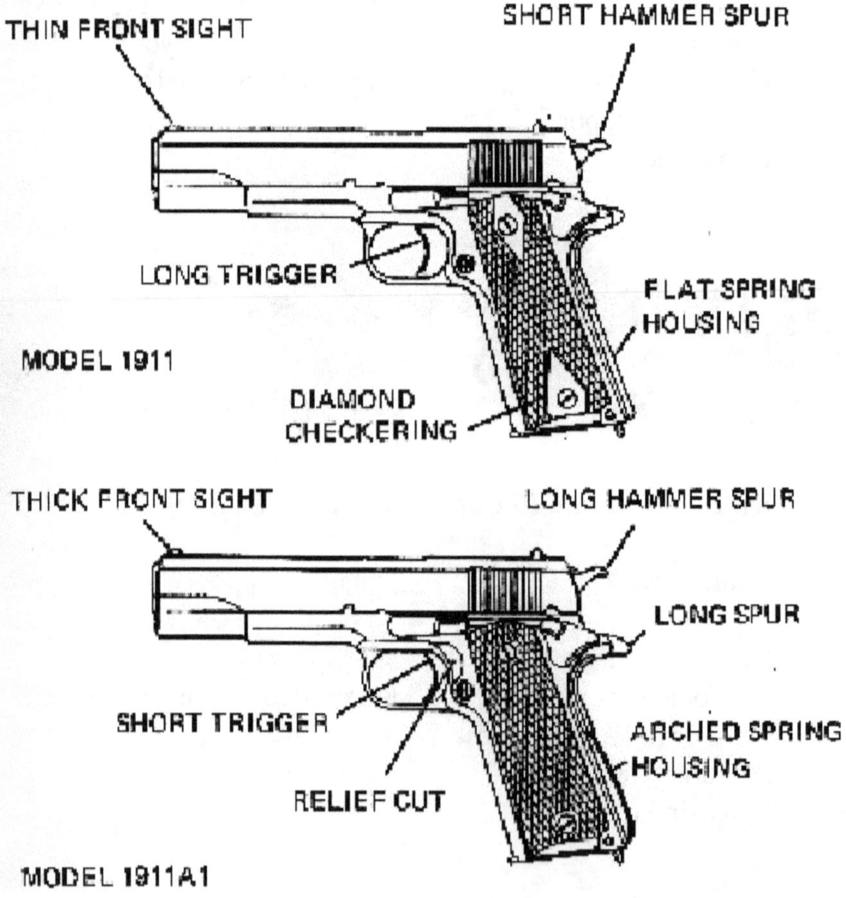

THIN FRONT SIGHT   SHORT HAMMER SPUR

LONG TRIGGER   FLAT SPRING
                                    HOUSING

MODEL 1911

                    DIAMOND
                    CHECKERING

THICK FRONT SIGHT   LONG HAMMER SPUR

                                    LONG SPUR

SHORT TRIGGER   ARCHED SPRING
                                    HOUSING
          RELIEF CUT

MODEL 1911A1

Figure 76. Pistol Cal .45 M1911 and M1911A1.

Characteristics:	Data:
Recoil operated	Maximum range (1,500 meters)
Semi-automatic	Max. Effective range (50 meters)
Magazine Fed	
Air cooled	
Hand weapon	

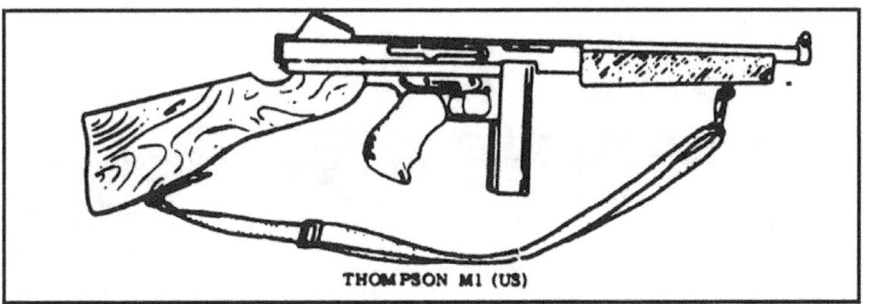

Figure 77. Thompson Submachine Gun M1A1 Cal .45.

Characteristics:	Data:
Air cooled	Cyclic rate of fire (600-725 rpm)
Blowback operated	Max. effective range (100m)
Semi or fully automatic	Maximum range (1,500 meters)
Shoulder weapon	
Magazine fed	

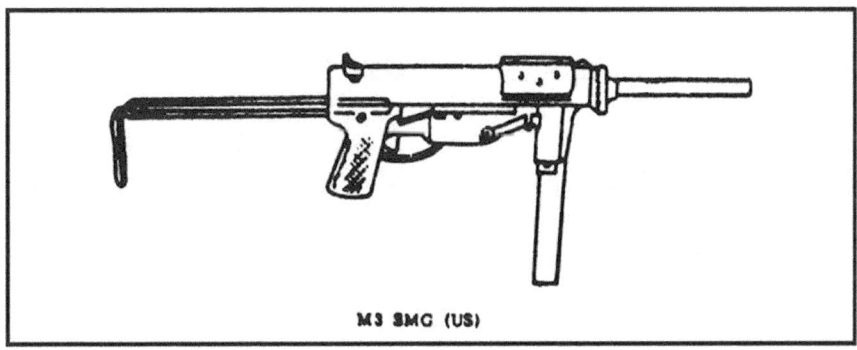

Figure 78. Submachine Gun M-3.

Characteristics:	Data:
Air cooled	Maximum range (1700 yds)
Blowback operated	Max. effective range (100 yds)
Automatic Shoulder weapon	
Magazine fed	

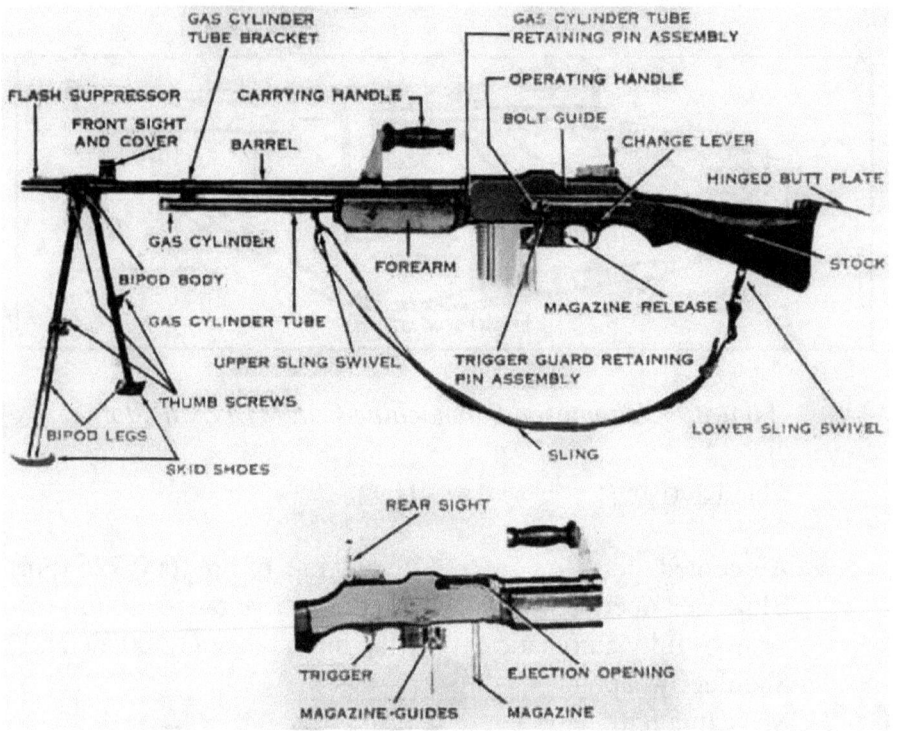

Figure 79. Browning Automatic Rifle M1918A2.

Characteristics:	Data:
Air cooled	Maximum range (3,500 yds)
Magazine fed	Max. effective range (500 yds)
Shoulder weapon	
Gas operated	
Fully automatic	
Data:	

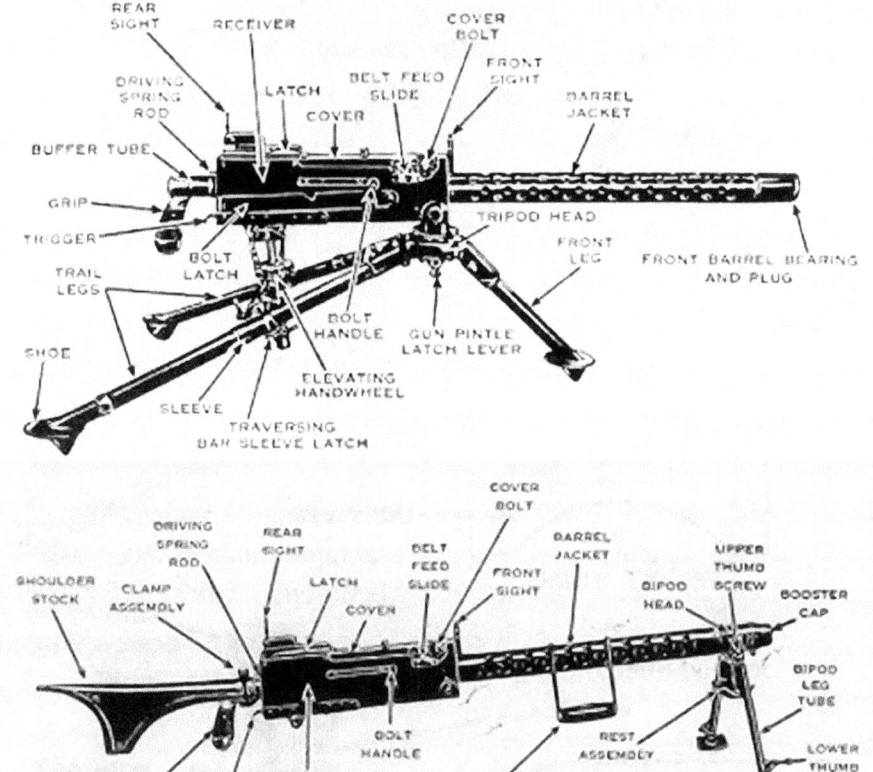

*Figure 80. Browning Machine Gun Cal .30 M1919A2 on M3 mount
(top) and M1919A6 on bipod (bottom).*

Characteristics:	Data:
Belt-fed	Max. effective range (1,200 yds)
Recoil operated	Max. range (3,500 yds)
Air cooled	Max. rate of fire (600-675 rpm)
Fully automatic	Max. eff. rate of fire (150 rpm)

Figure 81. Browning Machine Gun Cal .50, M2 HB.

Characteristics:	Data:
Air cooled	Max. effective range (2,000 yds)
Recoil operated	Maximum range (7,400 yds)
Fully and semi-automatic	Maximum rate of fire (500 rpm)
Alternate feed (right and left)	
Belt fed (metallic link)	

Figure 82. 57 mm Recoilless Rifle M18A1.

Characteristics:	Data:
Air cooled	Maximum range (4800 yds)
Recoilless	Max. effective range (1900 yds)
Shoulder or mounted weapons	
Single-loaded	Bursting area (10 x 34 yds - HE)
Fires fixed ammunition	(17 yds radius - WP)

Safety:

The danger zone from back blast is triangular in shape. It extends approximately 50 feet to the rear of the point of emplacement and at its widest point covers a space of 20 feet on either side of the axis of the emplaced rifle. Do not face the weapon within 100 feet of the rear of its breech because of the danger of flying particles thrown up by the blast action. The following danger zone will be for all training:

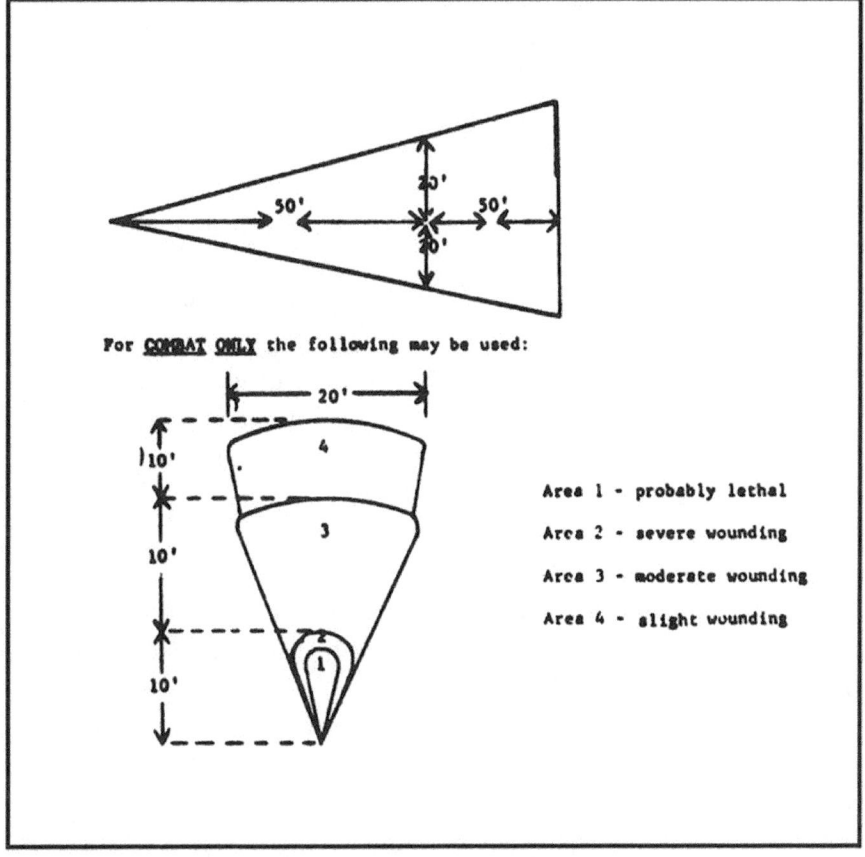

Figure 83.

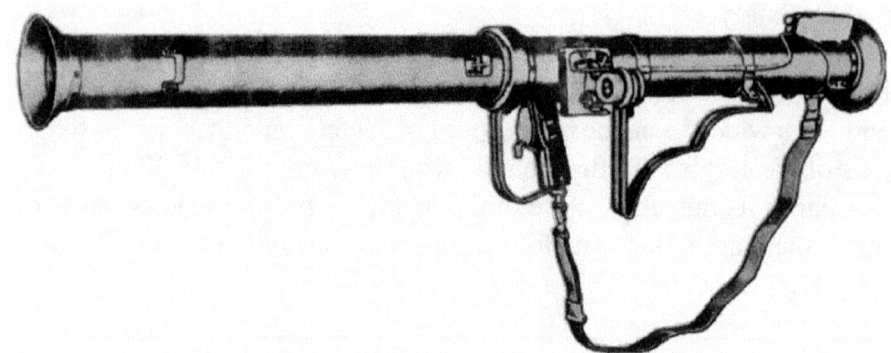

Figure 84. Rocket Launcher 3.5-Inch M20A1B1.

Characteristics:	Data:
Air cooled	Max. range (approx) (900 yds)
Smooth bore	Maximum effective range (Moving—200 yds; Stationary—300 yds)
Open tube (2 pieces)	Armor penet. (approx) (11 in)
Recoilless	Max. rate of fire (12-18 rpm)
Shoulder weapon	Sustained rate of fire (4 rpm)
Electrical firing mechanism	Burst. area approx (10 ×12 yds) (HEAT)

Safety precautions:

All loading and unloading are done on the firing line with the launcher on the gunner's shoulder. The muzzle is pointed down range, not toward the ground.

Face protection: For temperature below 70°F, the field protective mask must be used. For temperatures above 70°F, the anti-flash mask must be worn.

The weapon being of the recoilless principle has a danger zone to the rear. It is triangular in shape and consists of three zones. Before firing a rocket, clear the area to the rear of the launcher of personnel, material, and, dry vegetation as indicated in zone A & B.

Clear zone A, the blast area, of all prsonnel, ammunition, materials, and inflammables such as dry vegetation. The danger in this zone is from the blast of flame to the rear. Clear zone B of personnel and material unless protected by adequate shelter. The principle danger in zone B is from the rearward flight of nozzle clasure and/or igniter wires. An additional safety factor for training is contained in zone C.

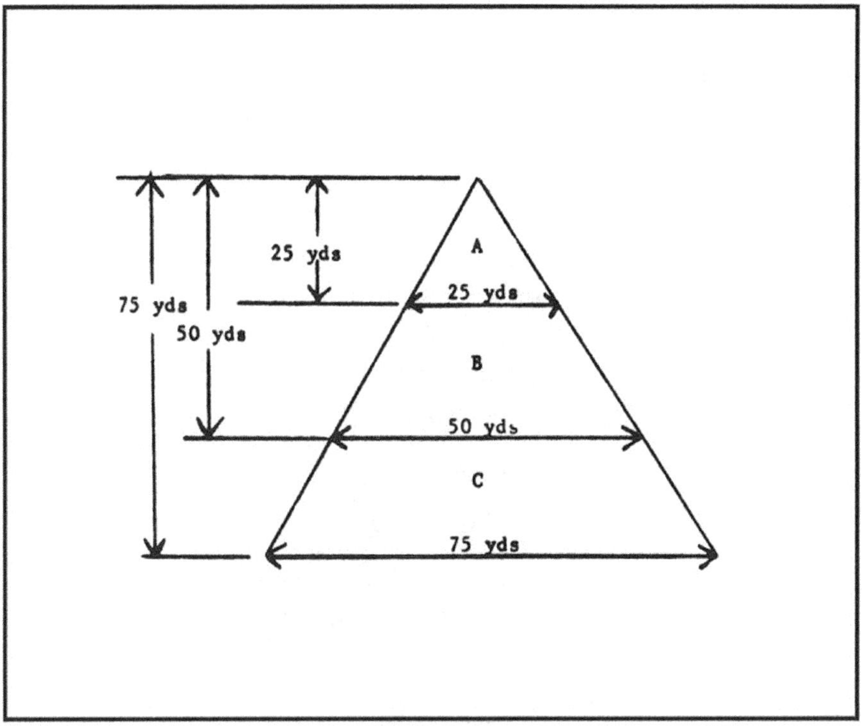

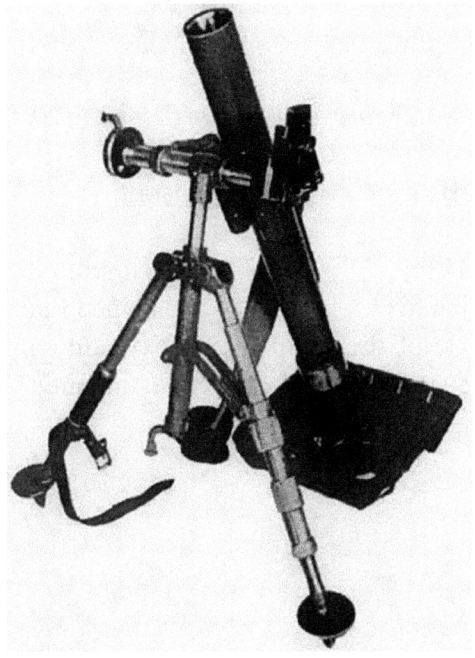

A

25 yds

25 yds

75 yds

50 yds

B

50 yds

C

75 yds

Figure 85.

Figure 86. Mortar 60 mm M-19.

Characteristics:	Data:
Smooth bore	Maximum rate of fire (30 rpm)
Muzzle loaded	Sustained rate of fire (18 rpm)
High angle-of-fire weapon	Bursting area (111 yd radius (HE and WP))

Figure 87. Mortar 81 mm M29.

Characteristics:	Data:
Smooth bore	Maximum rate of fire (24 rpm)
Muzzle loaded	Sustained rate of fire (3 rpm)
High angle-of-fire weapon	Maximum range (4,000 yds)
Drop fire	Bursting area (30 × 20 yds)

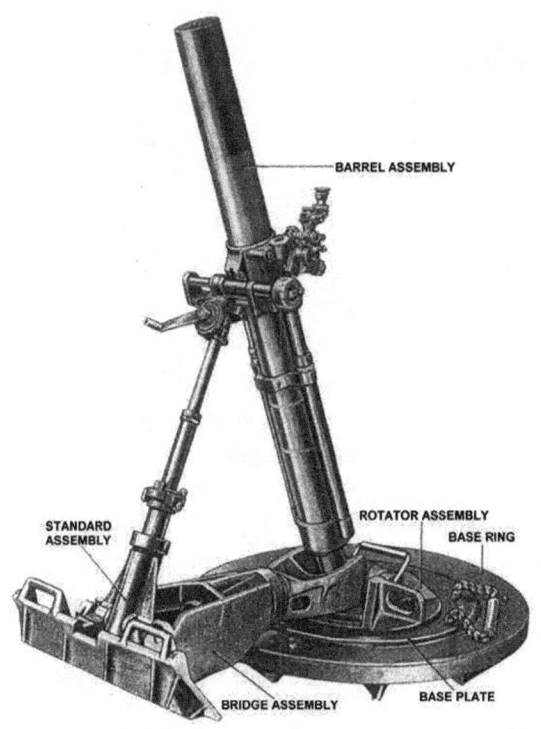

Figure 88. 4.2" Mortar M30.

Technical Data Characteristics:

Maximum range   6000 yds or 5500 m
Muzzle velocity   960 fps
Type of ammunition   HE, ILL and CHEM
Rate of fire   20 per min prolonged fire

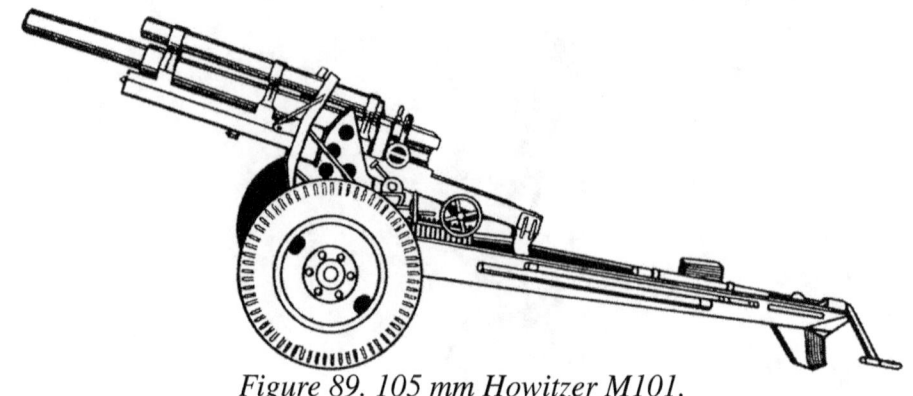

Figure 89. 105 mm Howitzer M101.

Technical Data and Characteristics:

Maximum range	11,270 meters
Muzzle velocity	1550 fps w/charge
Type of ammunition	HE, ILL, CHEM, HEAT,
Rate of fire	Rapid—4-8 per min
	Prolonged—100 rds per hr

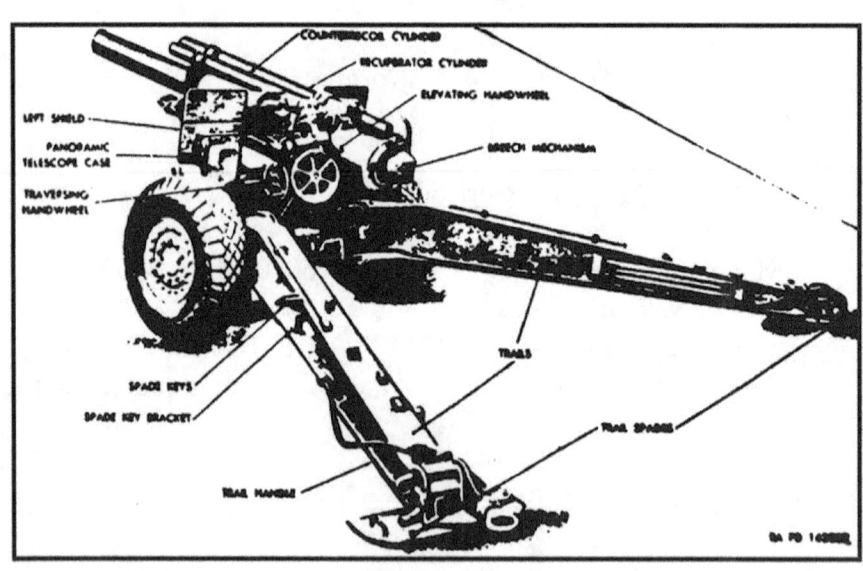

Figure 90. 155 mm Howitzer M114 or M114A1.

Technical Data and Characteristics:

Maximum range	14,955 meters
Muzzle velocity	1850 fps w/ch 7
Type of ammunition	HE, ILL and CHEM,
Rate of fire	Rapid fire—3 rds per min
	Prolonged fire—1 rd per min

I.   IMPROVISED RANGES.

Considerations.
Kind required by the training mission.
Travel time from camp to training area.
Security of training area.
Permission for use of area.
Safe impact area. (Clear before each firing.)
Terrain allows proper fields of fire for training to be conducted.
Vegetation in range area.
Materials available.
Labor and time available.

Shooting Gallery.
This is an introductory range to give the trainee practice in engaging a target with speed and accuracy.
Various targets such as bottles, plates, etc., of various colors and shapes are placed in clear view of the firer are at various angles from the firer. He is then instructed to engage targets by commands, giving direction and target. Example: "Right red can."
The firer is scored by number of hits and his speed in engaging the correct target.
Normally 3 seconds are allowed for each target; however, the instructor may vary this if the degree of training of his students so require.
Close Combat Range. Firer is put on firing line and targets are exposed for short periods of time. Firer engages target upon its appearance and is scored for bits and handling of weapon.

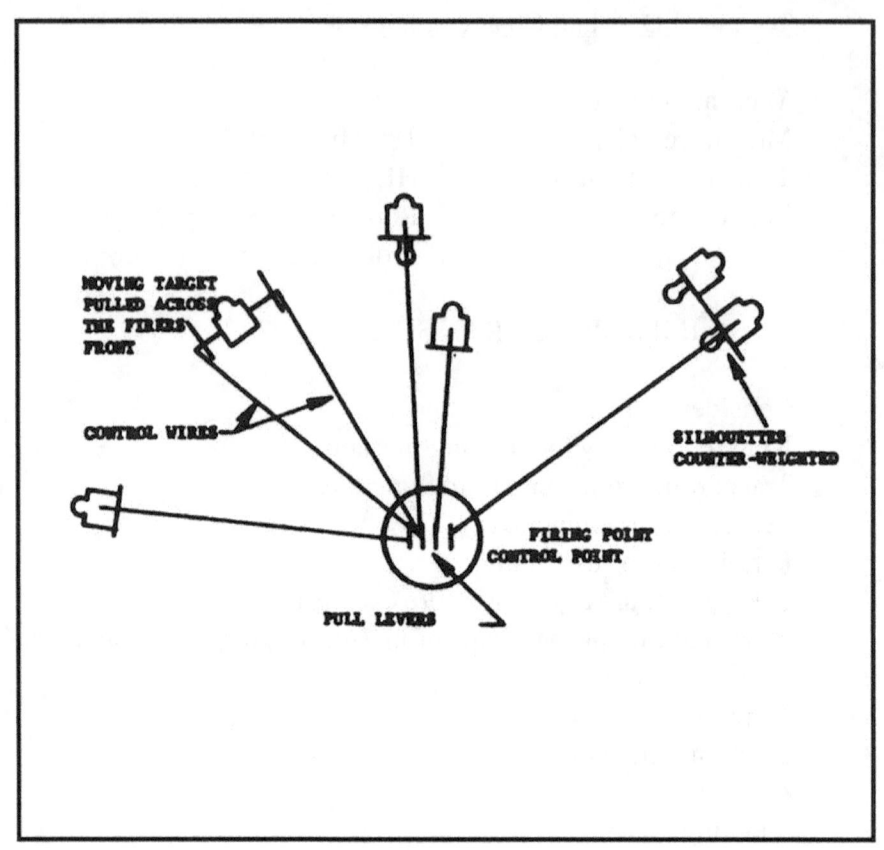

Figure 91.

KD Range:

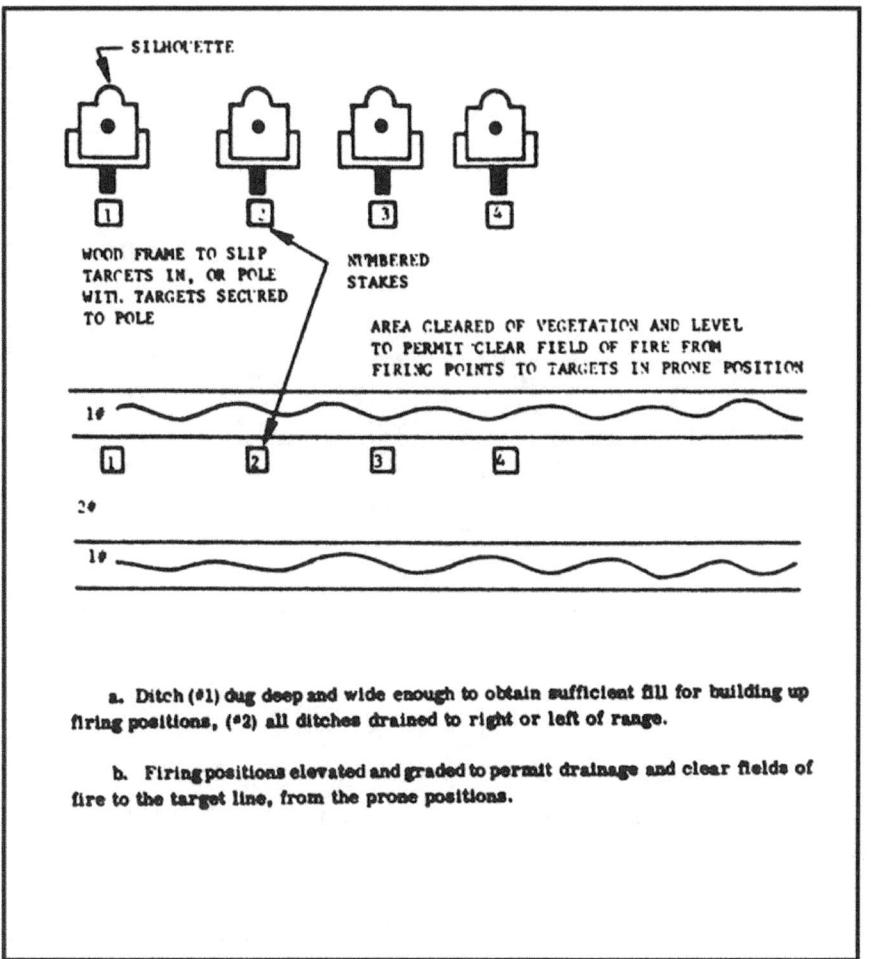

Figure 92.

Transition Range.

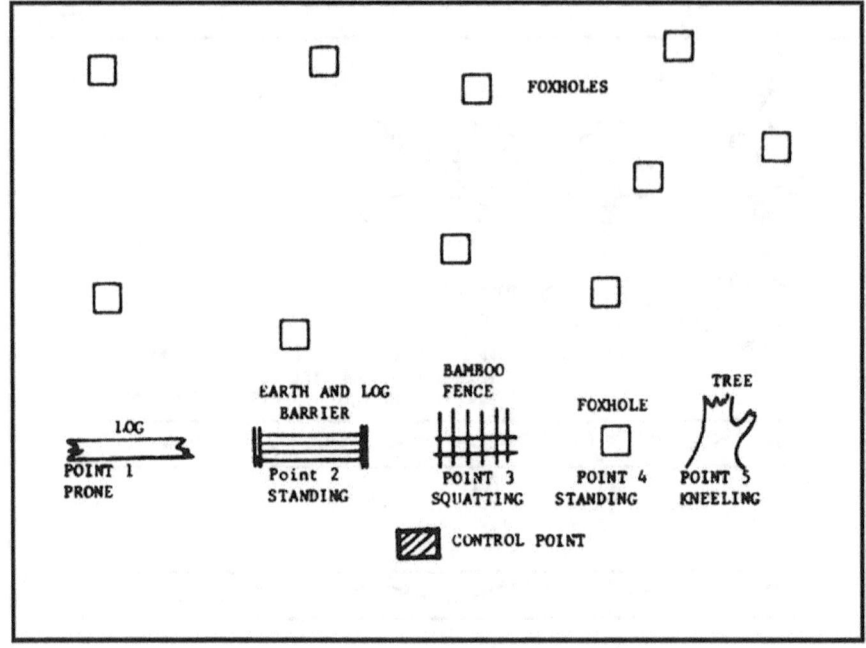

Figure 93.

Personnel are put into foxholes down range with silhouettes on poles. These foxholes must be dug deep enough to afford the operator protection. The range from firing point to target will be determined by your training program.

Personnel firing make up the designed position. When ready the range officer blows a whistle and all targets are exposed to the firer. The firer engages targets in his lane. After a designated time a signal is given and targets are lowered. All targets when hit will be lowered immediately.

Scoring may be accomplished by allowing so many points for each target hit and so many points for each unexpended round.

Jungle Lane.

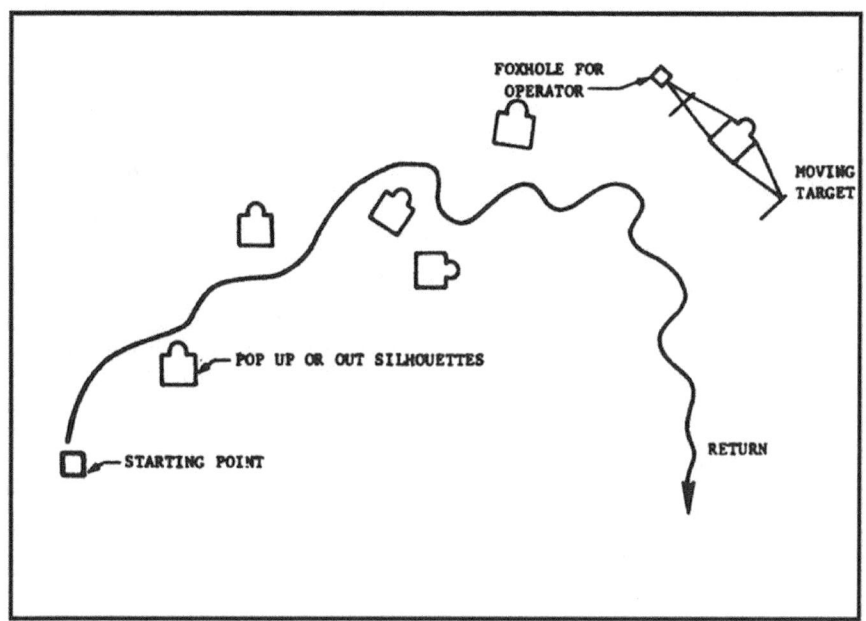

Figure 94.

This range may be used by individual firers or a small patrol. Targets are placed so they become exposed as the trainee rounds a bend or passes a thicket, etc. He will engage the target as soon as he observes it. Trainee is scored on his detection, accuracy, and handling of weapon.

Immediate Action Range.

This range may be employed for either vehicle or foot IA drill.

A path or road is selected with one or more good ambush sites on it. At one or more of these sites at least two foxholes are dug to accomodate two personnel each. These must be camouflaged from the trail or road. A silhouette target on a pole and an automatic weapon is placed in each foxhole. Additional targets which cannot be observed from road or trail but will be observed as the training unit deploys may be placed.

The trainees, organized in squad or larger units, are directed down the trail or road. When the instructor desires to trigger the ambush the automatic weapons in the foxholes open fire into a safe impact area and the silhouette targets are raised. The training unit then deploys, using the desired IA drill, engaging the target with live fire.

The instructor must exercise various safety measures as designating zones of fire and limiting points for deploying units.

Ambush Range.

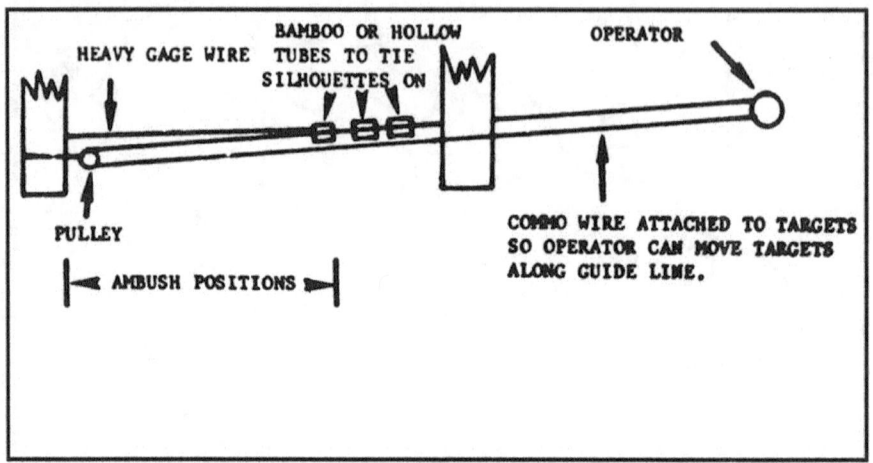

Figure 95.

This range should be built in vegetation such as the trainee will be operating in, with vegetation being left in place to make trainee select clear spaces to fire through.

A squad or similar size unit takes up an ambush position and the targets are then moved into the killing area. Targets are engaged at the ambush leader's signal.

Scoring can be accomplished by numbers of hits, triggering of ambush at most opportune time, distribution of fire and individual reaction.

II.   EXAMPLES OF TARGET CONSTRUCTION

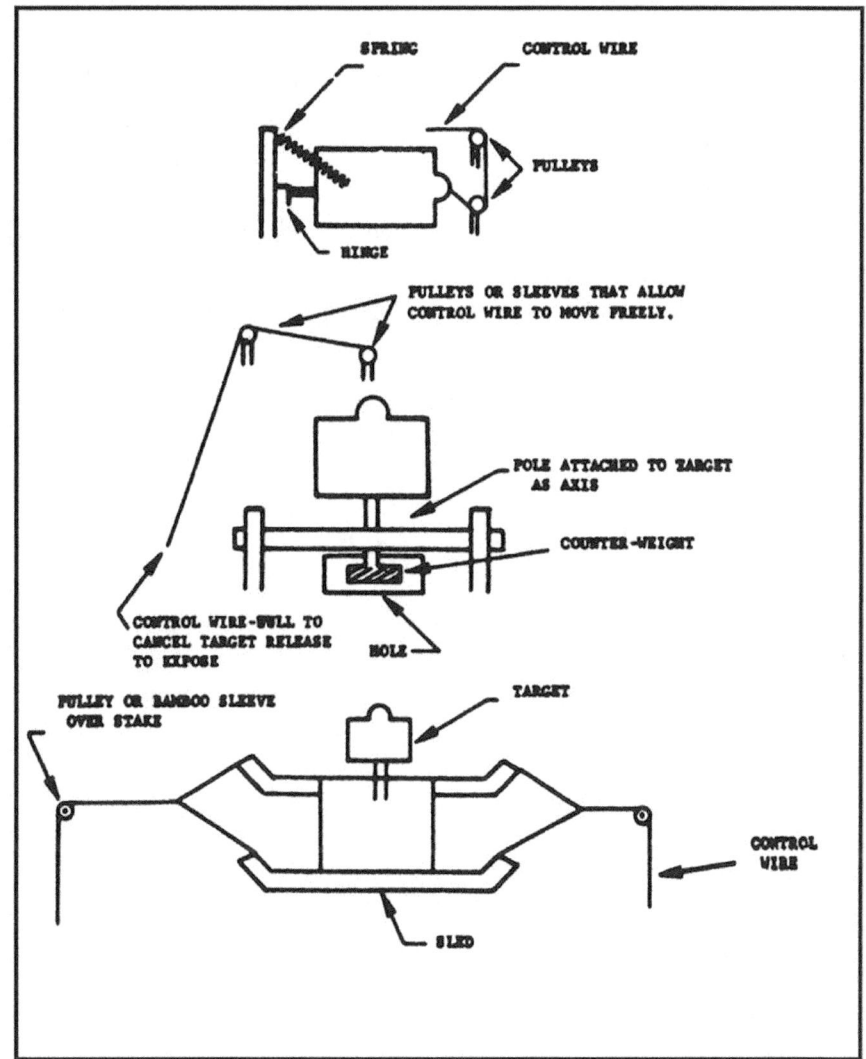

Figure 96.

III.   1000 INCH RANGE ZERO

To zero the rifle for 300 yards (battle sights), the shot group should be 1 1/4 inches above the point of aim at 1,000 inches.

This sight setting enables the soldier to hit his point of aim at a range of 300 yards.

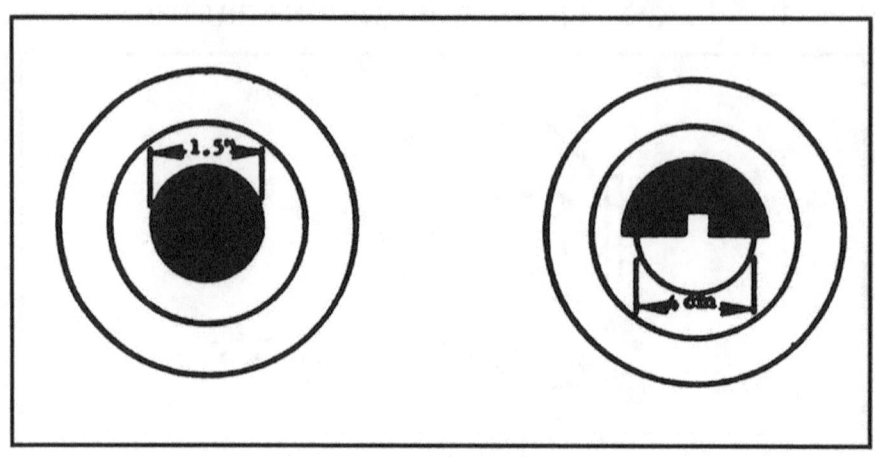

Figure 97. 1,500 inch target and 25 meter target.

IV.   25 METER RANGE ZERO.

To zero the rifle for 250 meters (battle sight), the shot group should be at the point of aim at 25 meters.

This sight setting enables the soldier to hit his point of aim at a range of 250 meters.

V.   WIND FORMULA.

To determine the clicks for full wind:

R (Range to target in hundreds yds) times V (wind velocity MPH) 15 (constant factor)

VI.   WORM FORMULA (any unit of measurement)

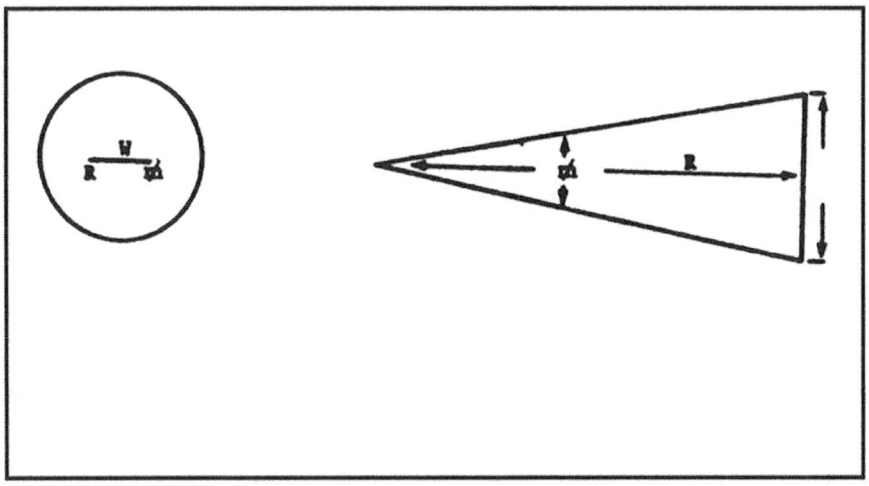

Figure 98.

One mil equals one at a range of one thousand.
W = Width To find W multiply R times R = Range in thousands To find R divide W by = Mils To find divide W by R

VII.   WIND VELOCITY CHART.

Degrees of angle (flag, handkerchief, etc)   MPH Wind

Degrees of angle	MPH Wind
15	5
30	10
50	15
60	20
70	25
80	30

Rule of Thumb Wind Velocity Formula. Hold paper, dust, or grass at arm's length and let it drop. Point to where it lands. Divide the angle between the arm extended and the body by four to get the MPH wind velocity.

CHAPTER 6
COMMUNICATION

PHONETIC ALPHABET AND INTERNATIONAL MORSE CODE

Letter	Word	CW Code	Letter	Word	CW Code
A	ALFA	·—	N	NOVEMBER	—·
B	BRAVO	—···	O	OSCAR	———
C	CHARLIE	—·—·	P	PAPA	·——·
D	DELTA	—··	Q	QUEBEC	——·—
E	ECHO	·	R	ROMEO	·—·
F	FOXTROT	··—·	S	SIERRA	···
G	GOLF	——·	T	TANGO	—
H	HOTEL	····	U	UNIFORM	··—
I	INDIA	··	V	VICTOR	···—
J	JULIETT	·———	W	WHISKEY	·——
K	KILO	—·—	X	XRAY	—··—
L	LIMA	·—··	Y	YANKEE	—·——
M	MIKE	——	Z	ZULU	——··

NUMBERS

1	·————	4	····—	7	——···	0	—————
2	··———	5	·····	8	———··		
3	···——	6	—····	9	————·		

PROWORDS AND PROSIGNS

PROWORD	PROSIGN	DEFINITION
THIS IS	DE	This transmission is from the station whose designation immediately follows.
OVER	K	This is the end of my transmission and a response is necessary. Go ahead; transmit.
OUT	AR	This is the end of my transmission and no answer is required. (Since OVER and OUT have opposite meanings, they are never used together.)
ROGER	R	I have received your last transmission satisfactorily.
SAY AGAIN	IMI	Repeat all of your last transmission.
I SPELL		I shall spell the next word phonetically.
CORRECTION	EEEEEEEE	An error has been made in this transmission. Transmission will continue with the last word correctly transmitted.
MESSAGE FOLLOWS		A message which requires recording is about to follow.

PROWORD	PROSIGN	DEFINITION
WILCO		I have received your message, understand it and will comply. (To be used only by the addressee. Since the meaning of the proword ROGER is included in that of WILCO, the two prowords are never used together.)
I SAY AGAIN	IMI	I am repeating transmission of portion indicated.
BREAK	BT	I hereby indicate the separation of the text from other portions of the message.
TIME		That which immediately follows is the time or date/time group of the message.
WAIT	AS	I must pause for a few seconds.
WAIT OUT	AS AR	I must pause longer than a few seconds.
GROUPS	GR	This message contains the number of groups indicated by the numeral following.
READ BACK	G	Repeat this entire transmission back to me exactly as received.
I READ BACK		The following is my response to your instructions to read back.
THAT IS CORRECT	C	What you have transmitted is correct.

PROWORD	PROSIGN	DEFINITION
WRONG		Your last transmission was incorrect. The correct version is _____.
RELAY (TO)	T	Transmit this message to all addressees or to the address designations immediately following.
ALL AFTER	AA	The portion of the message to which I have reference is all that which follows _____.
ALL BEFORE	AB	The portion of the message to which I have reference is all that precedes _____.
FROM	FM	The originator of this message is indicated by the address designation immediately following.
TO	TO	The addressee(s) whose designation (s) immediately following are to take action on this message.
SPEAK SLOWER		Reduce speed of transmission.
WORDS TWICE		Transmit(ting) each phrase (or each code group) twice.
VERIFY	J	Verify message (or portion indicated) with the originator and send correct version. (To be used only at the discretion of or by the addressee to which the questioned message was directed.)

PROWORD	PROSIGN	DEFINITION
I VERIFY		That which follows has been verified at your request and is repeated. (To be used only as a repy to VERIFY.)
SILENCE		"Silence" spoken three times means "Cease Transmission Immediately." Silence will be maintained until instructed to resume. Transmissions imposing "Listening" silence must be authenticated.
SILENCE LIFTED		Resume normal transmission. (Silence can be lifted only by the station imposing it or by a higher authority. When an authentication system is in force, transmission lifting "listening" silence must be authenticated.
SERVICE	SVC	The message that follows is a service message.
DO NOT ANSWER	F	Stations called are not to answer this call, receipt for this mcssage or otherwise to transmit in connection with this transmission. (When this proword is employed, the transmission shall be ended with the proword OUT.)

PROWORD	PROSIGN	DEFINITION
DISREGARD THIS TRANSMISSION	EEEEEEEE AR	This transmission is in error. Disregard it. (This proword shall not be used to cancel any message that has been completely transmitted and for which receipt or acknowledgment has been received.)
FLASH	Z	Precedence FLASH. (Reserved for initial enemy contact reports or special emergency operational combat traffic.)
IMMEDIATE	O	Precedence OPERATIONAL IMMEDIATE. (Reserved for important TACTICAL messages pertaining to the operation in progress.)
PRIORITY	P	Precedence PRIORITY. (Reserved for important messages which must have precedence over routine traffic.)
ROUTINE	R	Precedence ROUTINE. (Reserved for all types of message which are not of sufficient urgency to justify higher precedence, but must be delivered to the addressee without delay.)

PROWORD	PROSIGN	DEFINITION
FIGURES		Numerals or numbers follow. (Optional)
EXEMPT	XMT	The addressee designation immediately following are exempted from the collective call.
INFO	INFO	The addressee designation immediately following are addressed for information.
UNKNOWN STATION		The identity of the station with whom I am attempting to establish communication is unknown.
GROUP NO COUNT	GRNC	The groups in this message have not been counted.
EXECUTE	IX 5 Sec Dash	Carry out the purport of the message or signal to which this applies. (To be used only with the executive method.)
EXECUTE TO FOLLOW		Action on the message or signal which follows is to be carried out upon receipt of the proword "EXECUTE." (To be used only with the executive method.)

OPERATING SIGNALS

QRA	Station Name
QRK	Readability
QRL	Are you busy
QRM	I am being interfered with
QRN	I am troubled with static

OPERATING SIGNALS

QRQ	Send faster
QRS	Send slower
QRU	Nothing for you
QRV	Ready
QRX	I will call again at
QRZ	You are called by
QSA	Signal strength
QSB	Signals fading
QSD	Your key is defective
QSL	Acknowledge receipt
QSV	Send V's
QSY	Change transmitting frequency
QSZ	Send groups twice
QSW	I am going to transmit on frequency
QTB	Check your group count
ZBO	Message for you
ZKB	Take control of net until
ZKE	Reporting into net
ZKJ	Close down until
ZUE	Affirmative
ZUG	Negative
ZUH	Unable to comply
ZUJ	Stand by
ZXU	Unable to decipher
ZXV	Check encipherment

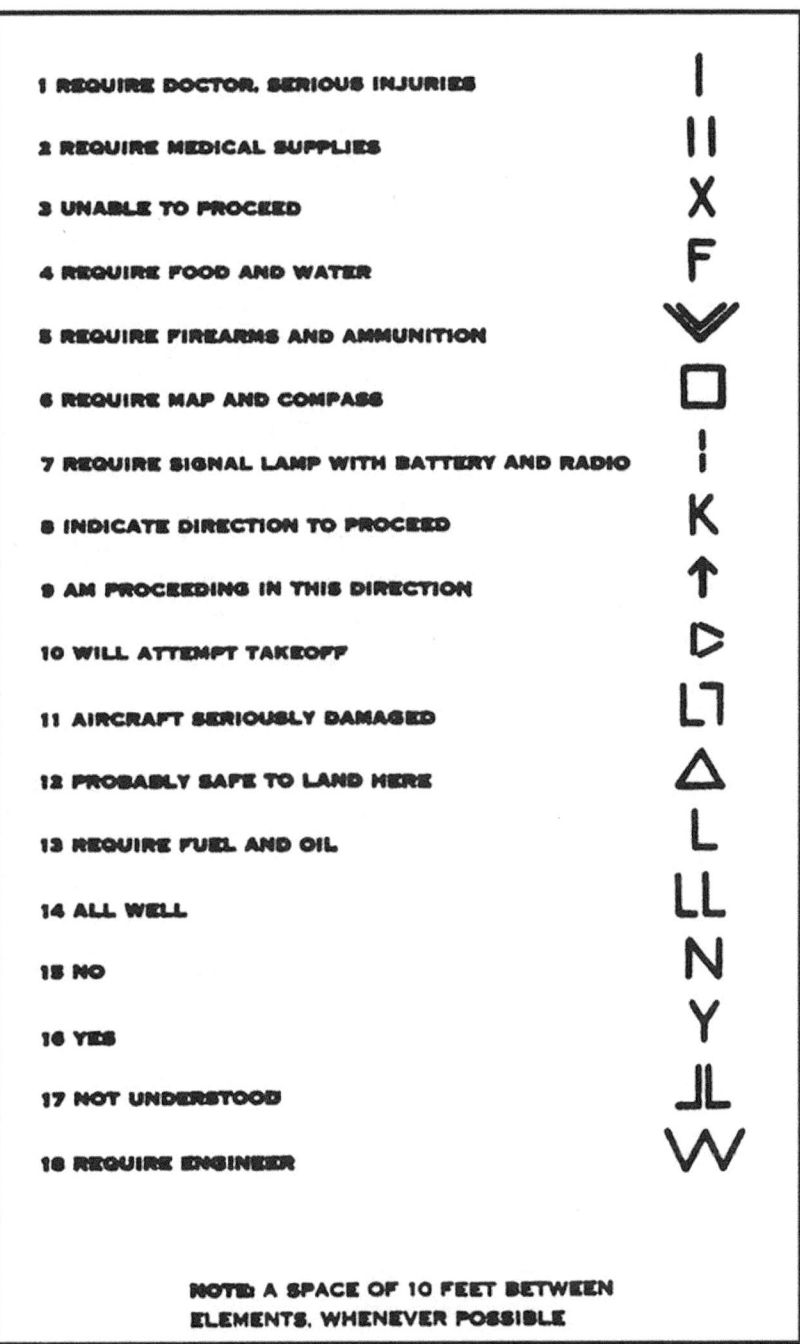

1 REQUIRE DOCTOR, SERIOUS INJURIES

2 REQUIRE MEDICAL SUPPLIES

3 UNABLE TO PROCEED

4 REQUIRE FOOD AND WATER

5 REQUIRE FIREARMS AND AMMUNITION

6 REQUIRE MAP AND COMPASS

7 REQUIRE SIGNAL LAMP WITH BATTERY AND RADIO

8 INDICATE DIRECTION TO PROCEED

9 AM PROCEEDING IN THIS DIRECTION

10 WILL ATTEMPT TAKEOFF

11 AIRCRAFT SERIOUSLY DAMAGED

12 PROBABLY SAFE TO LAND HERE

13 REQUIRE FUEL AND OIL

14 ALL WELL

15 NO

16 YES

17 NOT UNDERSTOOD

18 REQUIRE ENGINEER

NOTE: A SPACE OF 10 FEET BETWEEN
ELEMENTS, WHENEVER POSSIBLE

Figure 99. Ground-Air Emergency Code.

I.   COMMUNICATOR'S CHECK LISTS.

a.   Radio:

    1.   High Ground (FM).
    2.   Clearing with no obstructions (FM).
    3.   Antenna oriented with receiving station, clear of obstructions (AM).
    4.   Radio camouflaged properly.
    5.   Radio set properly grounded (AM).
    6.   Security around radio site.
    7.   Transmitting site moved, using around the clock method, but not going in a circle (AM).
    9.   Antenna properly loaded (AM).
    10.   Transmitter and Receiver on proper frequency.
    11.   Secret documents not at radio site.
    12.   Message encrypted currently (AM).
    13.   Radio site sterlized after departure.

b.   Visual Signals:

    1.   Signals properly placed so as to be easily read.
    2.   Signals simple and brief.
    3.   Operator properly oriented on signals and procedures.
    4.   Signalling device within range of receiver's vision.
    5.   Signals not too obvious.
    6.   Alternate signals.

c.   Audio Signals:

    1.   Easily understandable.
    2.   Clear and loud.
    3.   Signals related to surrounding noises.
    4.   Signals changed frequently.
    5.   Signals simple and clear.
    6.   Alternate signals included.

d.   Message Center:

    1.   Message center established.
    2.   All incoming and outgoing messages logged.
    3.   Code (and alternate code) made up for internal use.
    4.   Encrypting checked before transmission.

II.   RADIO NETTING CONSIDERATIONS.

When planning a radio net certain technical factors must be considered in connection with the equipment available. They are:

Emission—Are the radios compatible?

Are all radios going to operate voice or CW?

Frequency—Can the radios operate within the same frequency band?

Modulation—AM works only with AM. FM only with FM.

Range—Do not plan a net beyond the transmission range of the weakest set.

Crystals—Are proper crystals on hand if needed?

Terrain—Are appropriate high points available for radio stations if line of sight communications are planned?

Operating factors to consider are:

Schedule of operation.

Proficiency of operators.

Communications security.

Physical security of codes ciphers.

Cryptographic security and operating information (SOI).

Transmission Security.

AN/GRC - 100 - PREPARATION FOR OPERATION.

1.   Connect appropriate power supply to power source.

2.   Connect transmitter and receiver to appropriate power supply or source.

3.   Connect a lead 25 feet or less to the ground post of transmit ter and a good ground. (If good ground is not available utilize a counterpoise).

4.   Connect a lead from "RCVR ANT" on transmitter to "ant" on

131

receiver.

5.   Connect a lead from "RCVR GRD" on transmitter to "grd" on receiver.

6.   Connect antenna to "ANT" post on transmitter. Select proper length of antenna to correspond with operating frequency. Antenna must be at least one quarter wave length long. Radio Set AN/GRC - 109 will load properly on end fed single line antenna that is exactly 1/2 wave length or any multiple thereof. To provide for a better indication in the antenna load lamp, the physical length of the wire may be adjusted 10 percent.

7.   Set tuning dial on receiver to receiving frequency.

8.   Check tuning chart on front of transmitter and turn controls to the settings indicated.

9.   Tune all controls on transmitter in proper sequence for maximum glow on the indicator lamps. (Retune the first lamp slightly to prevent a chirping signal from being emitted.)

10.   Connect headset to terminals on the receiver and adjust gain for desired level.

11.   Tune the beat frequency oscillator control to the ON postion for CW reception and adjust for desired tone.

12.   Power Supplies (Must have power source): a. Large Power Supply PP-2684 (AC-DC). b. Small Power Supply PP-2685 (AC only). c. Voltage Regulator CN-690 (G-43/U only).

13.   Power Sources:

        a. AC voltage 75-260 VAC 40-400 ops (with PP-2684 or PP-2685).

        b. 6 volt wet cell battery (with PP-2684).

        c. Hand generator G-43/G (with PP-2684, CN-690, or direct to transmitter).

        d. Gas Generator AN/UGP~12 (with PP-2684 or PP-2685).

        e. Dry Battery BA-317 or BA-48 (direct to receiver).

AN/GRC-84, 87, AN/VRC-34, PREPARATION FOR OPERATION.

1.   OFF SEND STANDBY switch to STANDBY.

2.   PHONE CW NET CAL switch to CAL.

3.   PHONE MCW CW switch to PHONE.

4.   A.F. gain control fully clockwise to STOP.

5.   R.F. gain control fully counter clockwise (OFF).

6.   Band switch to appropriate band.

7. Turn receiver tuning control to crystal check point nearest desired frequency. Increase R.F. gain control slightly until signal is heard. Adjust receiver tuning control until zero beat is heard on the strongest beat note in the vicinity of the crystal check point. Keep R.F. gain control adjusted to the point where the beat note is just audible.

8. PHONE CW NET CAL switch to NET.

9. PHONE NCW CW switch to CW HI.

10. XTAL MO band switch to MO of appropriate band.

11. Refer to calibration chart, set transmitter tuning control to same frequency as now appears on the receiver dial.

12. OFF SEND STANDBY switch to STANDBY when using GN 58 and BA 317.

13. Adjust A.F. gain for the desired volume and turn R.F. gain to mid point.

14. Adjust OSC CAL control until zero beat is heard. (Do not close microphone or key while performing this step.) Power must be obtained at this time from the generator.

15. Refer to calibration chart and set transmitter tuning to desired operating frequency and look tuning control.

16. Set receiver tuning control to desired operating frequency and tune receiver for zero beat with transmitter. Look tuning con trol. Must obtain power from generator.

17. Set antenna selector control to the highest numbered position for the type of antenna being used. Close key or microphone and rotate the antenna tuning control until indicator glows and adjust for maximum glow.

18. Set receiver and transmitter switches for the desired type of transmission and reception.

19. The set is now ready for operation.

II.   INTERPOLATION.

1. A dial calibration chart appears on each AN/GRC 87.
2. Its purpose is to relate dial settings to transmitting frequencies.
3. The charts on each set are different.
4. The dial calibration chart will not give you the dial setting for unlisted frequencies ... you must interpolate to find it.
5. Steps in interpolation:
   a. Subtract the next lower frequency from the desired frequency.
   b. Find the difference between the dial readings just above and just below the desired frequency.
   c. Multiply the values obtained in these two steps.
   d. If in band 2 or 3, divide by 20. If in band 1, divide by 50.
   e. Add the results of step above to the dial setting for the next lower listed frequency. This is the correct dial setting for your desired frequency.

EXAMPLE: Desired frequency is 4487 kcs:
1. Subtract 4480 from 4487 = 7
2. Subtract 1471 from 1491 = 20
3. Multiply 7 by 20   = 140
4. Divide 140 by 20   = 7
5. Add 7 to 1471   = 1478 proper dial setting

Freq	≠ 100 kc	≠ 20 kc	≠ 40 kc	≠ 60 kc	≠ 80 kc
hard 2					
4300	2279	1362	1324	1347	1368
4400	1389	1410	1431	1451	1471
4500	1491	1511	1530	1550	1569

III.   ANTENNA CONSIDERATIONS.

One of the most critical aspects of reliable radio transmission and reception is the proper design, utilization and location of transmitting and receiving the antennas.

Antennas should be "cut" the wave length of the frequency being used. Most of the time, however, this is not practical, so a 1/2 or 1/4 wave length antenna is used.

The formulas below should be used to determine desired antenna lengths.

$$1/4 \text{ wave} = \frac{234}{F}$$ NOTE:

$$1/2 \text{ wave} = \frac{468}{F}$$ F is frequency in megacycles

$$1 \text{ wave} = \frac{936}{F}$$ Antenna Lengths are in feet

NOTE: When using radio set AN/GRC-109 with 1/2 wave length end fed antenna, the antenna may be adjusted by ±10 percent of the exact wavelength.

TABLE NR.IV. ANTENNA LENGTH CHART.

FREQUENCY MEGACYCLE	FULL WAVE LENGTH	FREQUENCY MEGACYCLE	FULL WAVE LENGTH	FREQUENCY MEGACYCLE	FULL LENGTH
1	936	21	44.6	41	22.8
2	468	22	42.6	42	22.2
3	312	23	40.6	43	21.8
4	234	24	39	44	21.2
5	187.2	25	37.4	45	20.8
6	156	26	36	46	20.4
7	133.6	27	34.6	47	19.8
8	117	28	33.4	48	19.4
9	104	29	32.2	49	19
10	93.6	30	31.2	50	18.8
11	85	31	30.2	51	18.4
12	78	32	29.2	52	18
13	72	33	28.4	53	17.6
14	66.8	34	27.6	54	17.4
15	62.4	35	26.8	55	17
16	58.4	36	26	56	16.8
17	55	37	25.2	57	16.4
18	52	38	24.6	58	16.2
19	49.2	39	24	59	15.8
20	46.8	40	23.4	60	15.6

LENGTHS ARE IN FEET

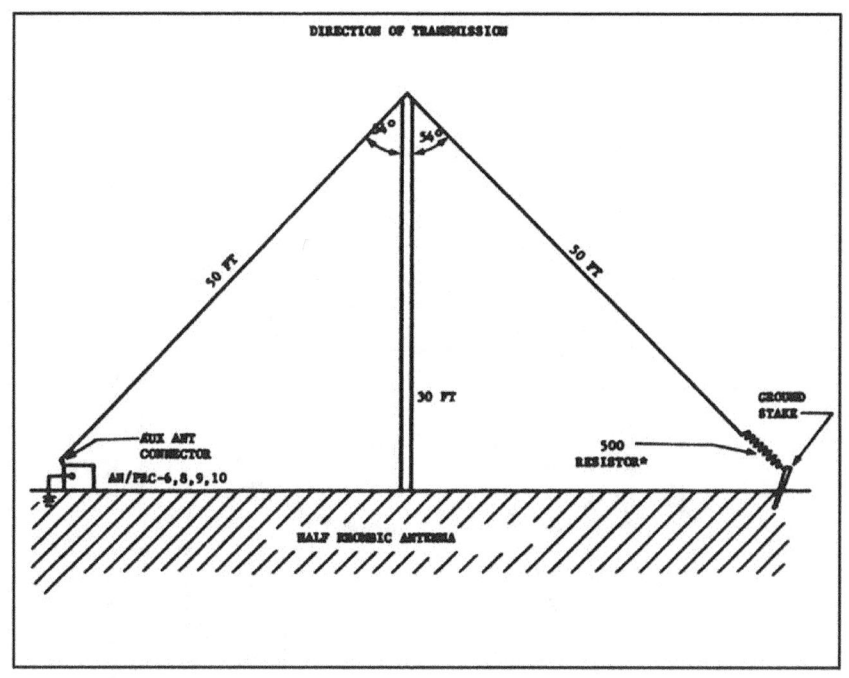

THIS ANTENNA IS BEST USED WITH FM RADIO.

*DESIRABLE, BUT NOT NECESSARY

Figure 100.

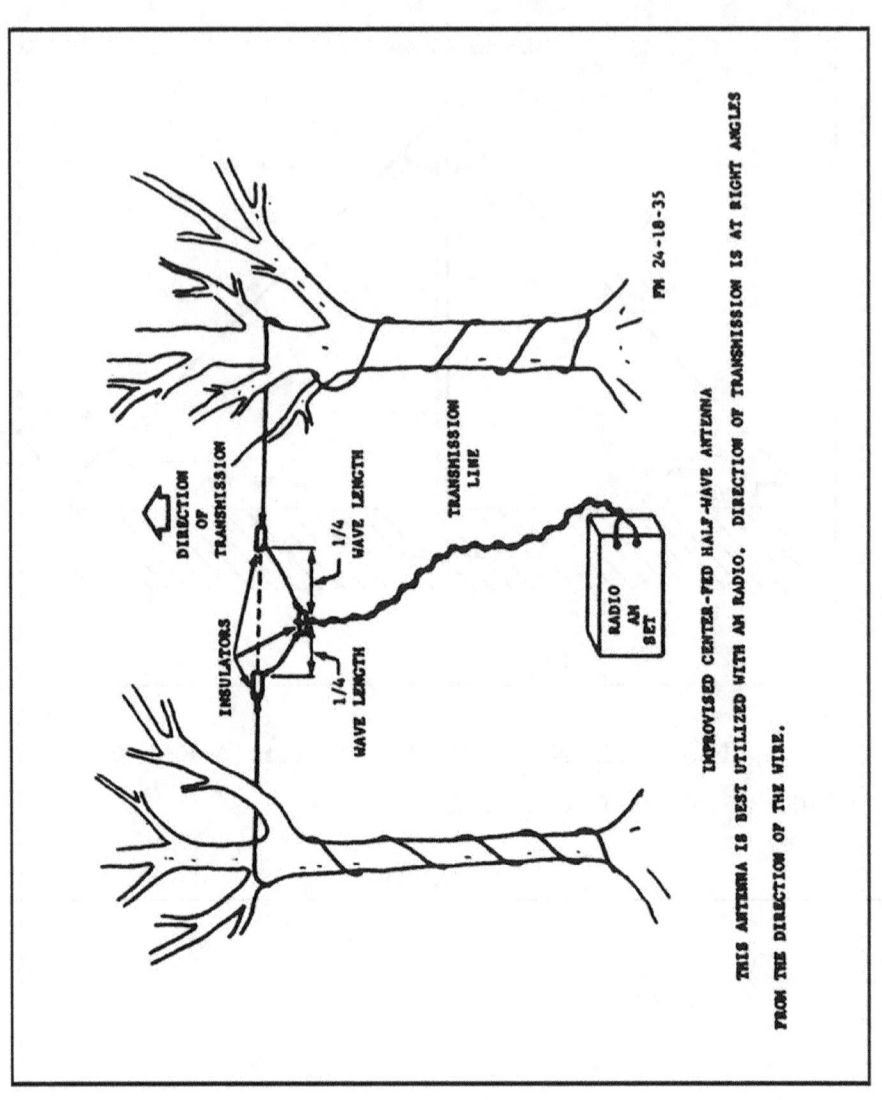

Figure 101.

138

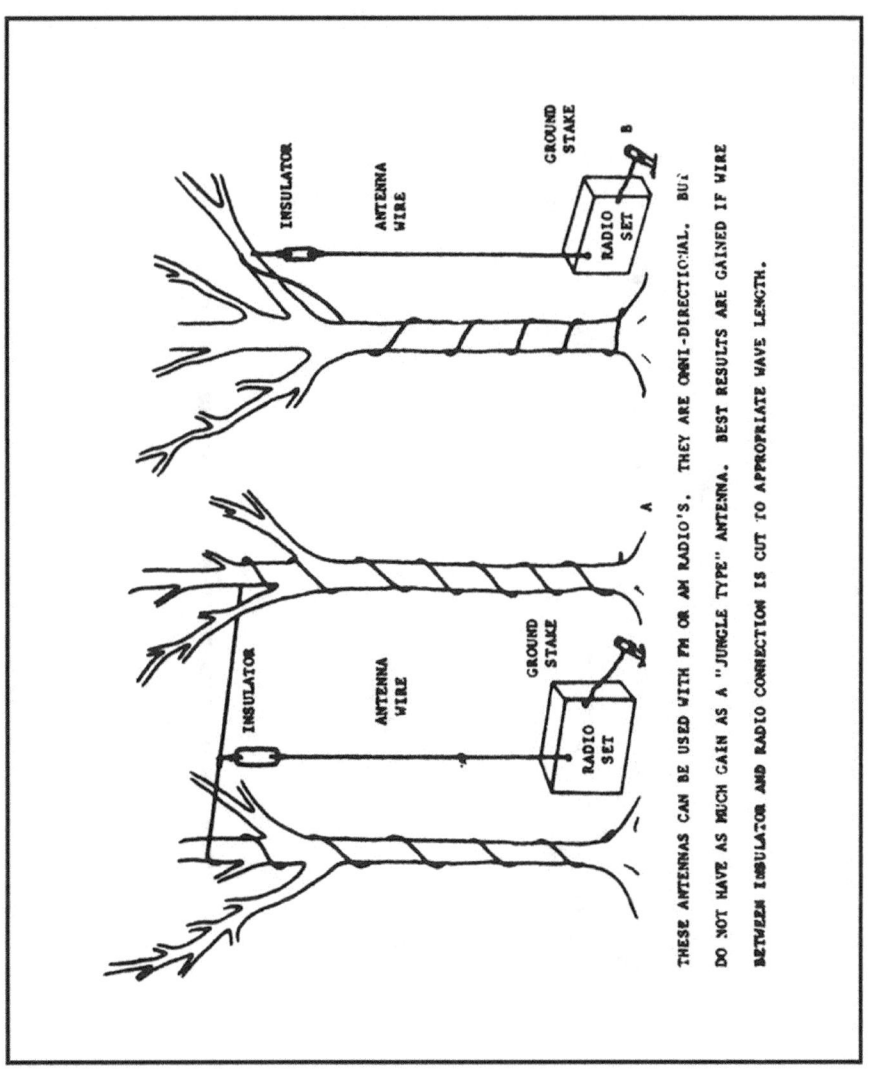

Figure 102.

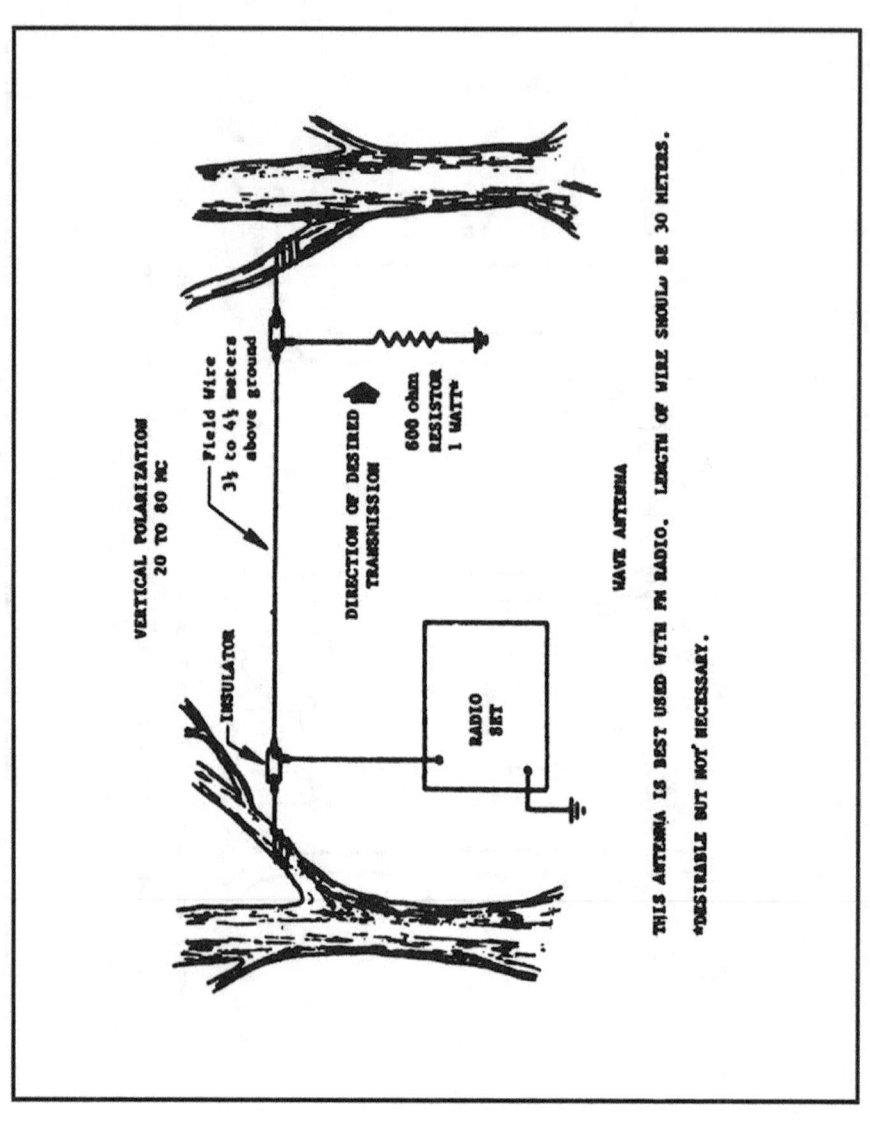

Figure 103.

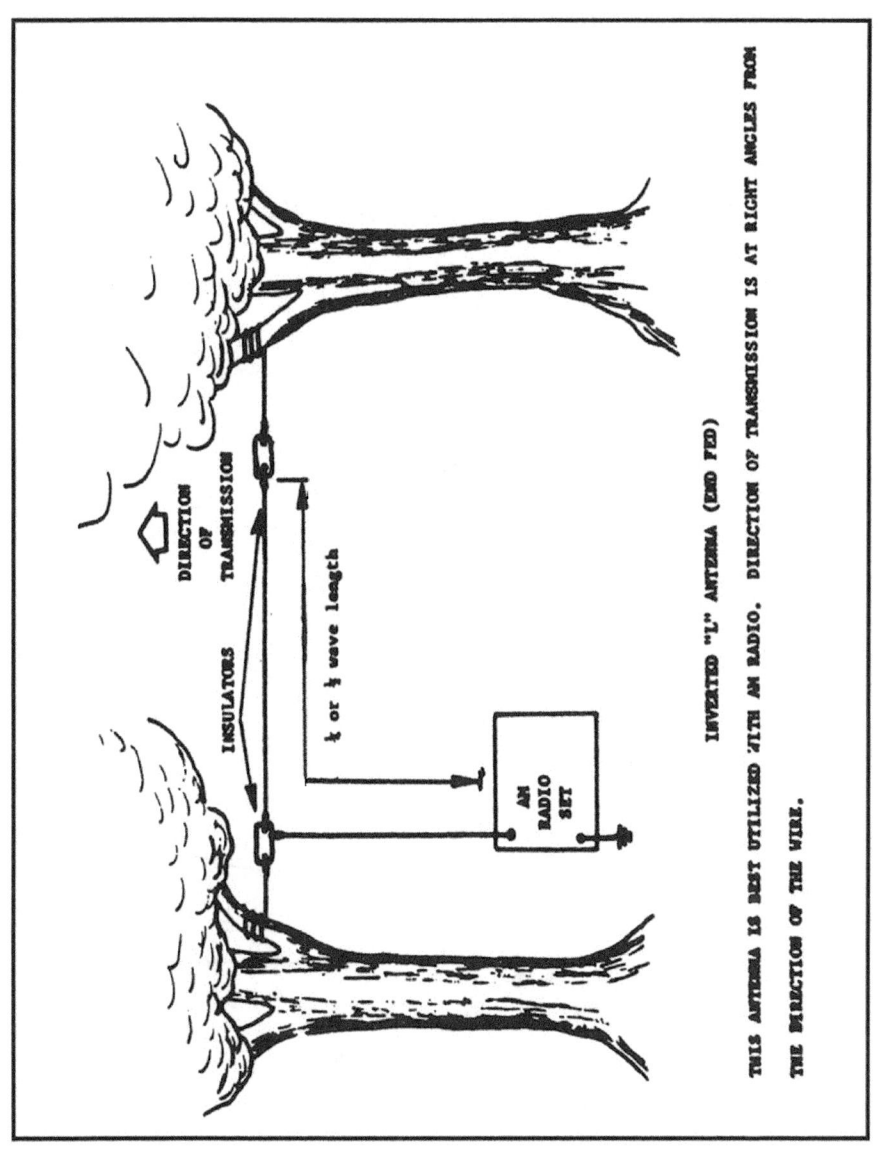

Figure 104.

141

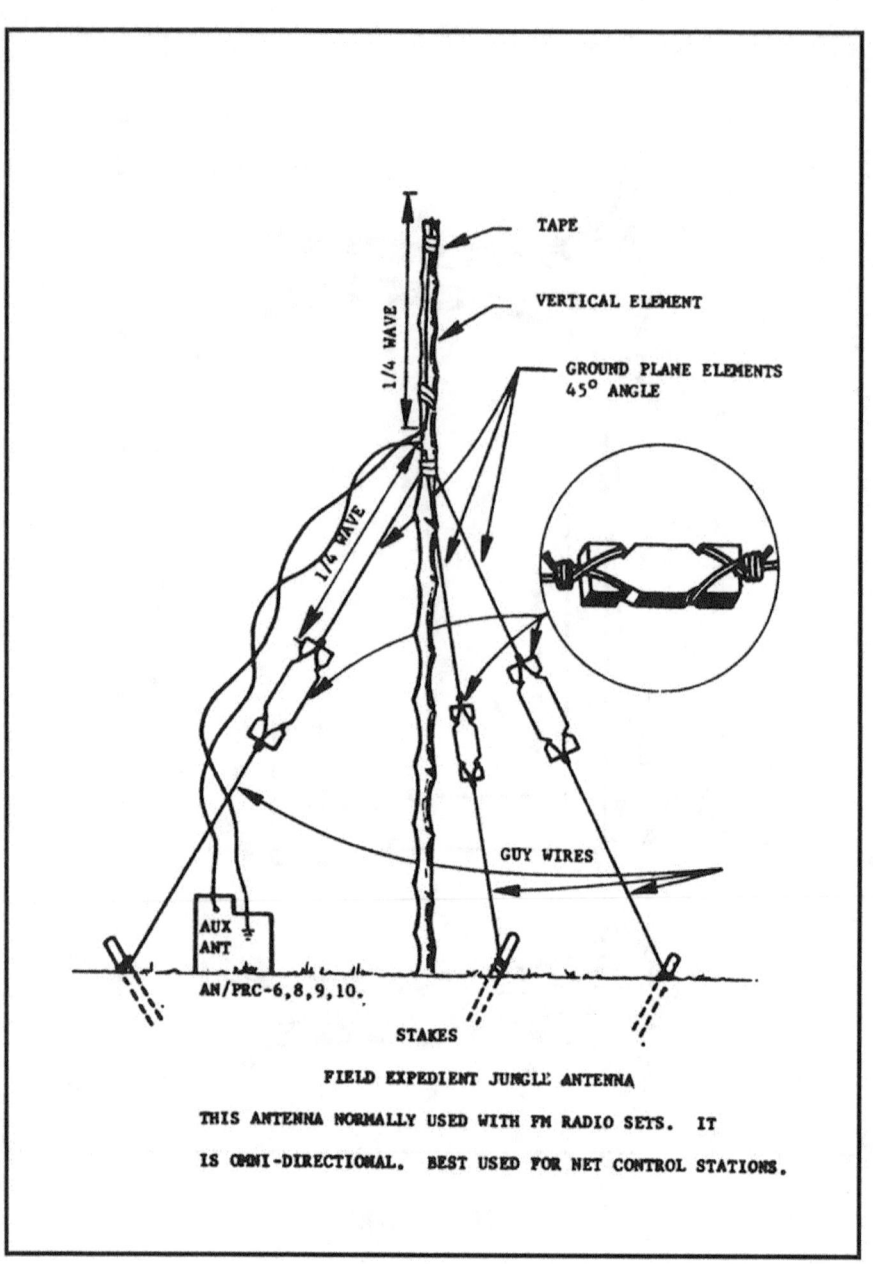

Figure 105.

142

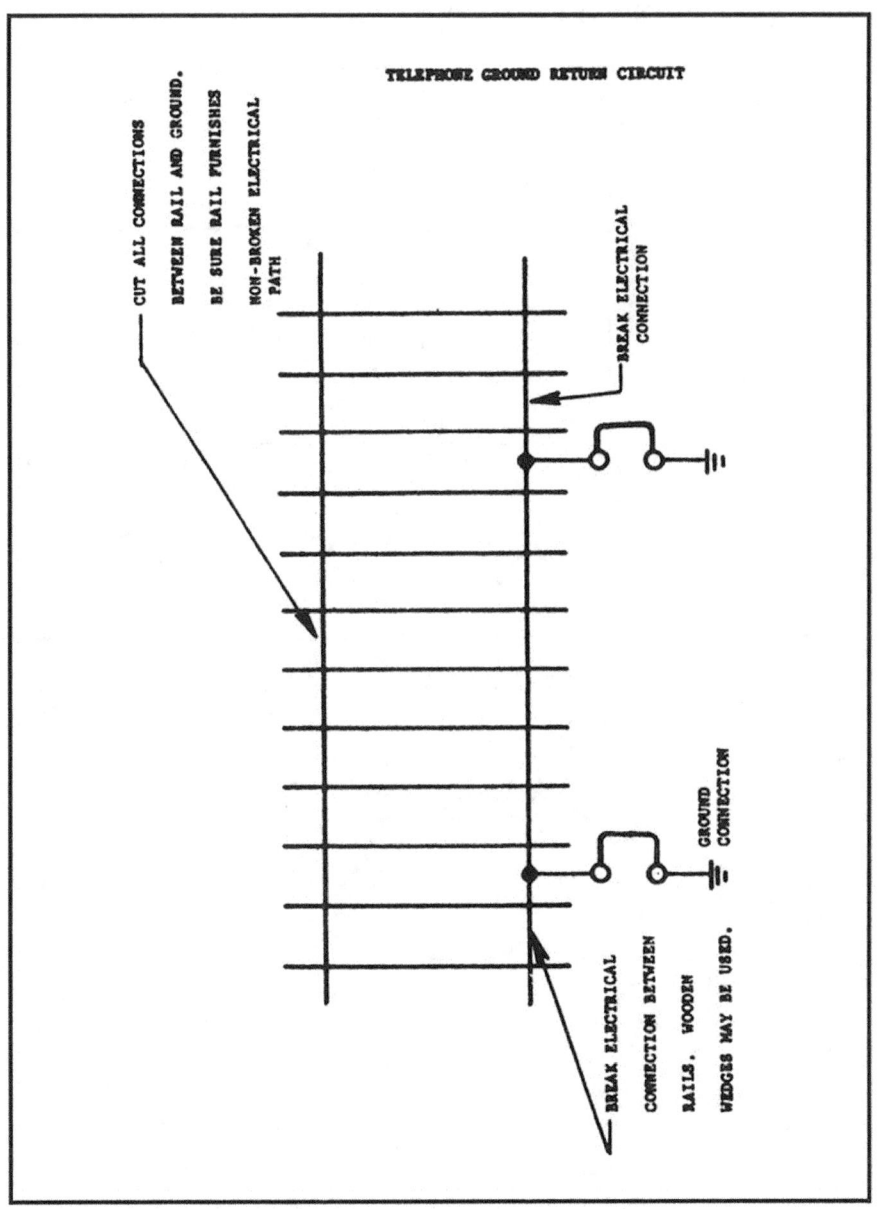

Figure 106.

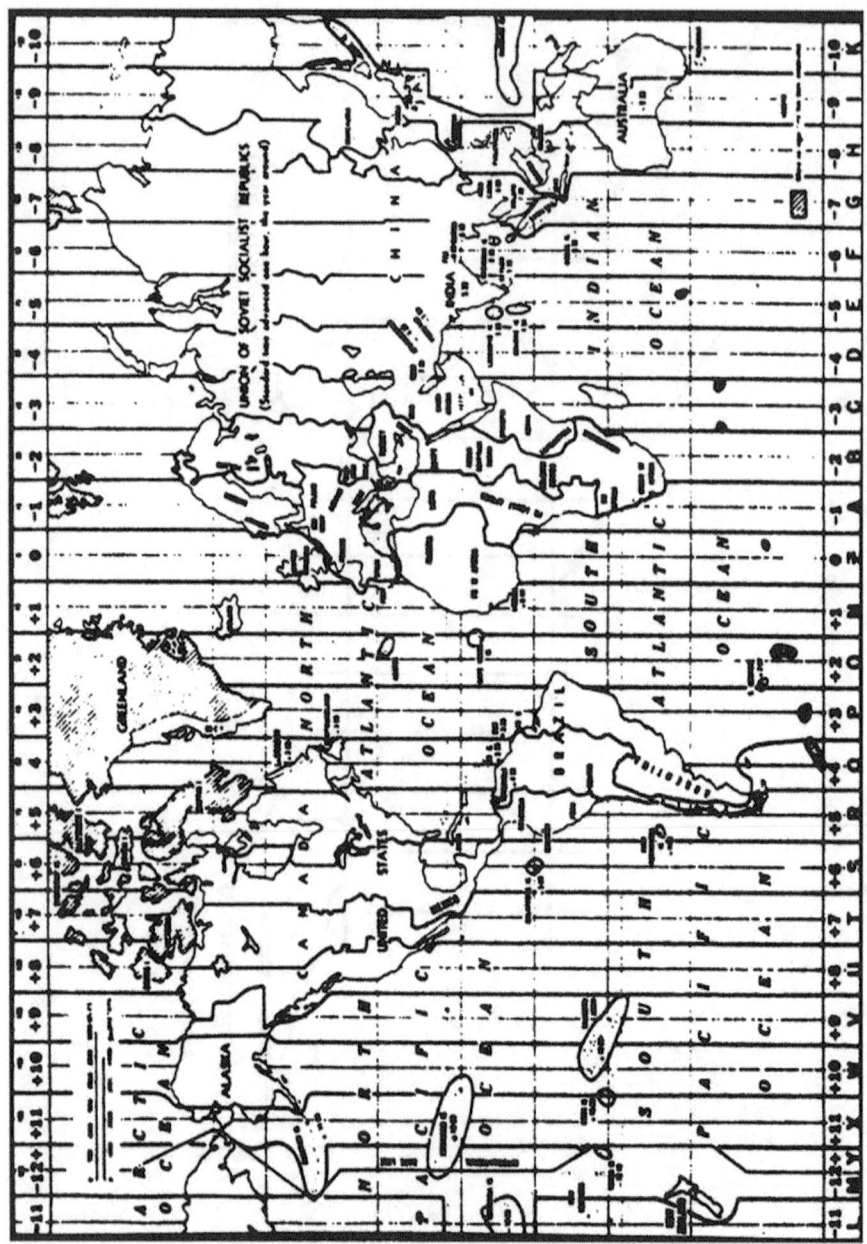

Figure 107. World Time Zone Map.

CHAPTER 7
FIRST AID

FIRST AID TREATMENT.

AILMENT	SYMPTOMS	TREATMENT
Shock	Pale face	Lay patient on back.
	Cold clammy skin	Lower head, elevate feet.
	Rapid weak pulse	Loosen clothing, keep warm.
	Shallow breathing	Feed hot liquids if conscious.
Wound		Expose wound. Control bleeding. Apply sterile dressing. Treat for shock.
Fracture	Pain and tenderness	Handle with care; splint before moving.
	Partial or complete loss of motion	Support the limb on either side until splint is applied.
	Deformity Swelling	Splints must be long enough to reach beyond joints above and below fracture and must be tied twice above and below break to immobilize limb.
	Discoloration	Pad all splints. Treat for shock.
Burn	First degree:	Carefully remove or cut clothing

		away from burned area.
	Skin red No blister	
	Second degree:	Don't open blisters.
	Skin blistered	Cover area with sterile dressing.
	Third degree: Skin destroyed and charred	Keep burned areas apart by separate bandages. Treat for shock.
Sunstroke (direct exposure to sun)	Flushed face	Remove from sun.
	Dry skin	Take off all clothing.
	Strong rapid pulse	Elevate head and shoulders. Apply cool compresses or bathe patient in cool water.
	Spots before eyes Headache High temperature	Give patient cool salt water.
Heat Exhaustion	Dizziness	Move patient to shade.
	Nausea	Treat as for shock.
	Pale face	Give cool salt water.
	Cramps Cold clammy skin Weak pulse	

Frostbite	Numbness	Do not rub, bend or expose to extreme heat or further cold.
	Waxy colorless tissue Stinging pain at onset	Warm area to body temperature by holding close to warm body or exposing to warmth no higher than 95 degrees.
Snake Bite	Bites from poisonous snakes will cause swelling in about 45 min.	Treat ALL snake bites as poisonous.

I. SNAKE BITE:

Apply a tourniquet 3/4 inch above the bite. Don't put it on too tightly. The object is to retard the flow of blood returning to the heart; not to cut off circulation altogether. Make sure there is a pulse below where the tourniquet is applied. DO NOT CUT. Leave this for the medical officer or medic. As swelling progresses up the limb, move the tourniquet, keeping it 2/3 inches ahead of the swelling.

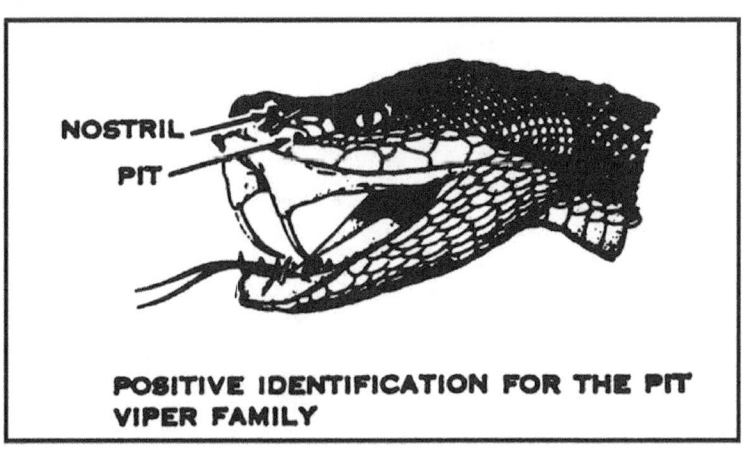

Figure 108. Identifying a pit viper.

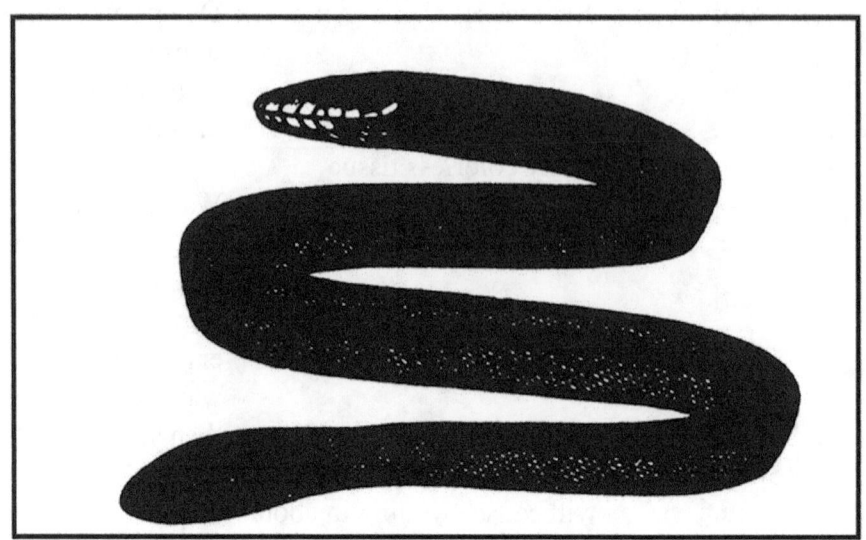

Figure 109. Sea Snake.

If the victim stops breathing, begin mouth-to-mouth rescuscitation and continue till a medical officer arrives.

Send for medical help ASAP. If no one else is available to go, and the patient is conscious and alert, it will be necessary to leave him alone while you go. Before leaving, give him instructions to remain still and to move the tourniquet as required.

Keep the patient QUIET. If it is impossible to bring help to him then carry him to aid.

Make tourniquet just tight enough to retard flow of lymph.

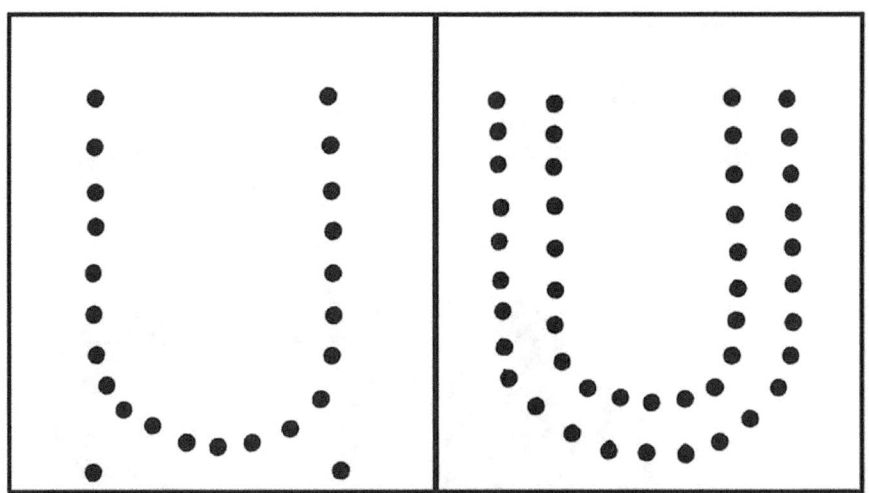

Teeth marks of poisonous snake (note fang marks).

Teeth marks of non-poisonous snakes (note two rows).

Figure 110.

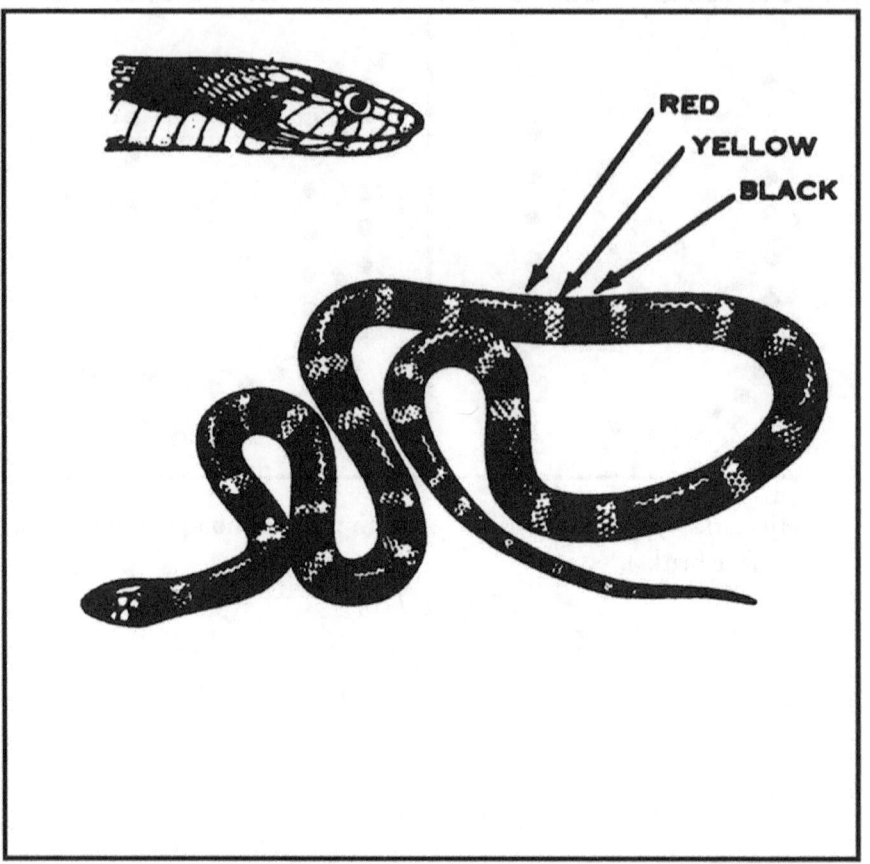

Figure 111. Coral Snake.

II.   FIRST AID PRINCIPLES:

a.   Stop bleeding.
b.   Protect the wound.
c.   Prevent or treat for shock.
d.   Splint fractures.

III.   CONTROL OF BLEEDING:

a.   Elevate injured member if not fractured.
b.   Apply pressure bandage.
c.   Use pressure points if blood is gushing (wherever strong pulse is felt). (See figure on pressure points.)
d.   Use tourniquet only as last resort.

IV.   PRESSURE POINTS:

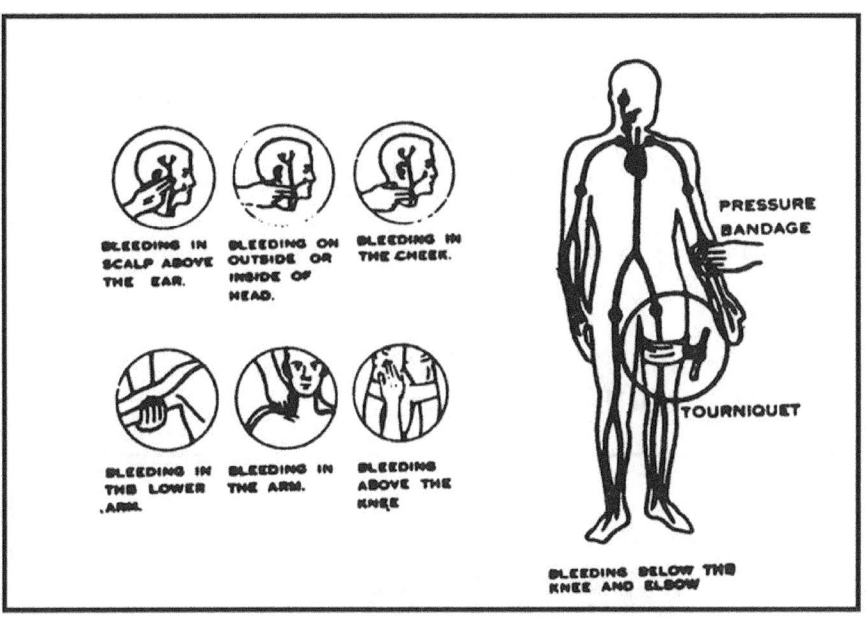

Figure 112.

V.   TYPES OF BLEEDING:

Arterial—spurting
Venous—flowing.
Capillary—oozing.

VI.   ARTIFICIAL RESPIRATION—BACK-PRESSURE
      ARM-LIFT:

Place your hands on the flat of the victims back so that the palms lie just below an imaginary line running between the arm pits. With tips of your thumbs just touching, spread you fingers downward and outward.

Rock forward, with elbows straight until your arms are almost upright and let the weight of the upper part of your body press slowly, steadily and evenly downward on your hands on the victim's back.

Release the pressure by removing the hands from the back

without a push and rock slowly backward on your heels. As you do this, slide your hands outward and grasp the victim's arms near the elbows.

Draw the victim's arms upward and toward you with just enough of a lift to feel resistance and tension at the victim's shoulders. Do not bend your elbows. Then lower his arms to the ground.

Continue this action until normal breathing is resumed by victim.

VII.   ARTIFICIAL RESPIRATION—MOUTH-TO-
        MOUTH:

Figure 113.

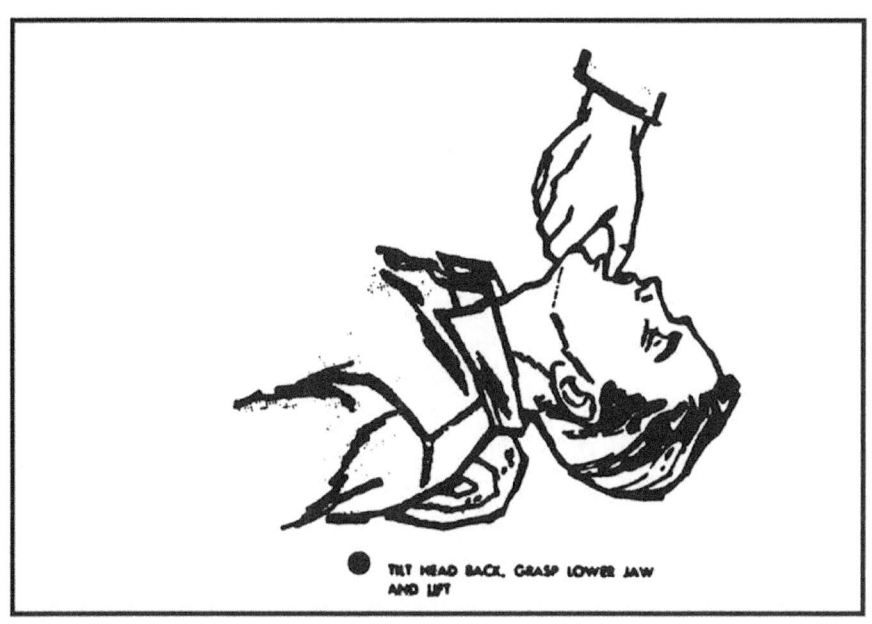

TILT HEAD BACK, GRASP LOWER JAW AND LIFT

Figure 114.

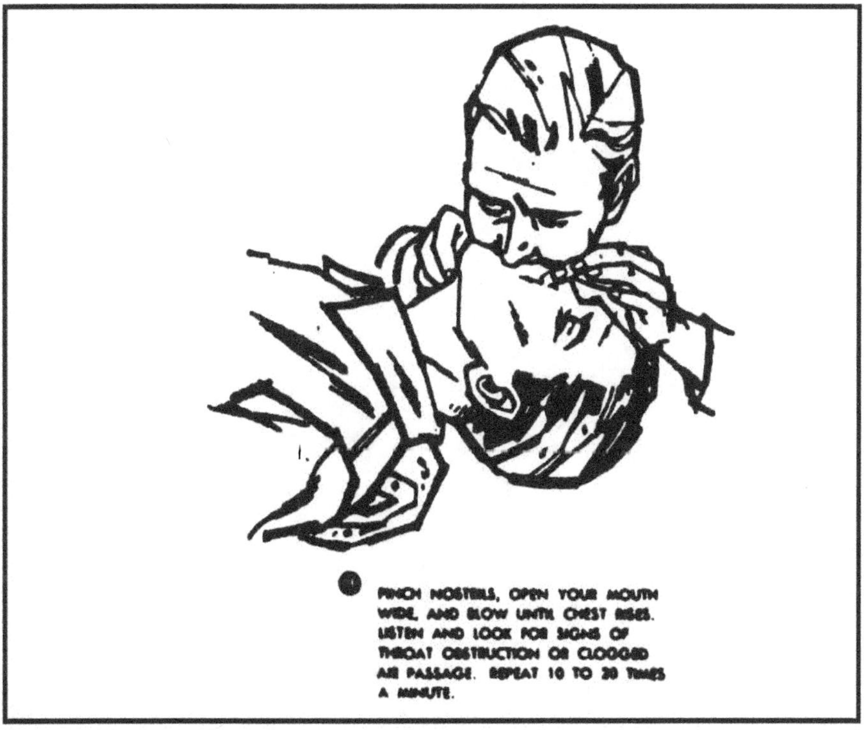

PINCH NOSTRILS, OPEN YOUR MOUTH WIDE, AND BLOW UNTIL CHEST RISES. LISTEN AND LOOK FOR SIGNS OF THROAT OBSTRUCTION OR CLOGGED AIR PASSAGE. REPEAT 10 TO 20 TIMES A MINUTE.

Figure 115.

153

CHAPTER 8
SURVIVAL

HOT-WET SURVIVAL INFORMATION

Be Alert

Be Wary of Strangers

Guide on Trails to Friendly Villages

Follow or Float on Waterways to Sea Coast

Food Grows in Fields Near Villages

Conceal All Evidence of Your Being in an Area

A Few Feet into Dense Jungle Will Hide You

Insect Repellent Applied to Fiber Makes Good Tinder

Boil or Treat All Water Used for Drinking or Washing

I.   EVASION.

First, get as far away as possible. Sometimes this may mean several miles; at other times, just a few yards. Plan your escape, do not run blindly. Use your head—there is no substitute for common sense. As soon as possible, sit down, think out your problem, recall what you learned in training.

Pinpoint your location as accurately as possible, using your compass, sun, map, known landmarks, etc. If your compass is broken or lost, remember that when facing the sunrise, north is to your left. The following methods can be used for determining direction.

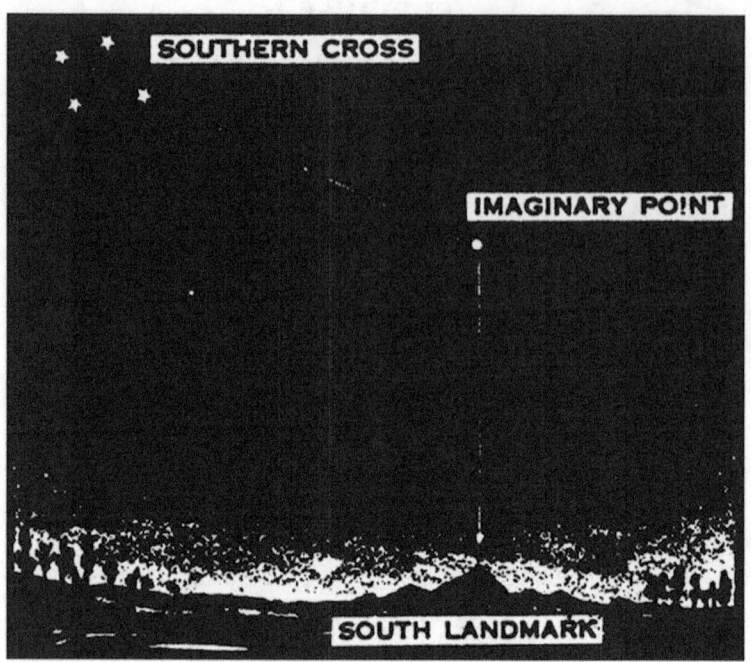

Figure 116. Southern Cross.

Using the Southern Cross: In the Southern Hemisphere you can find south by locating the Southern Cross. Compare this group of stars to a kite. If you can figure the length of the kite from tip to tail and extend an imaginary line from the tip of the tail four and one-half times the length of the kite, you can determine the approximate direction of south.

Using a watch to find north: General. The sun always appears to be south of the north temperate zone and north of the south temperate zone. A timepiece can be used to determine the direction of true north utilizing this fact, while compensating for the eastward to westward movement of the sun.

North temperate zone. Hold timepiece so that hour hand points at sun. Mentally draw an angle with its vertex at the center of the timepiece, one line passing through the number 12 and the other line along the hour hand.

Cut this angle in half and note its imaginary projections on the ground.

This imaginary line, bisecting the angle mentally drawn, points south, its reverse direction is north.

South temperate zone. a. Hold timepiece so the figure 12 points at the sun.

Mentally draw an angle with its center at the center of the timepiece and its sides passing through the figure 12 and along the hour hand.

Bisector now points north.

The diagrams below, graphically illustrate the methods of finding north described above.

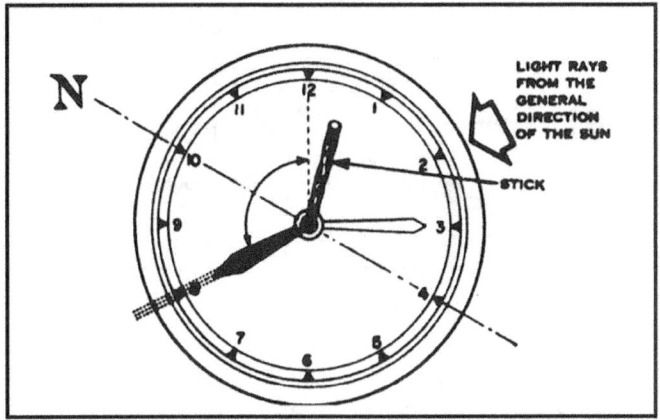

Figure 117. Finding north in the north temperate zone.

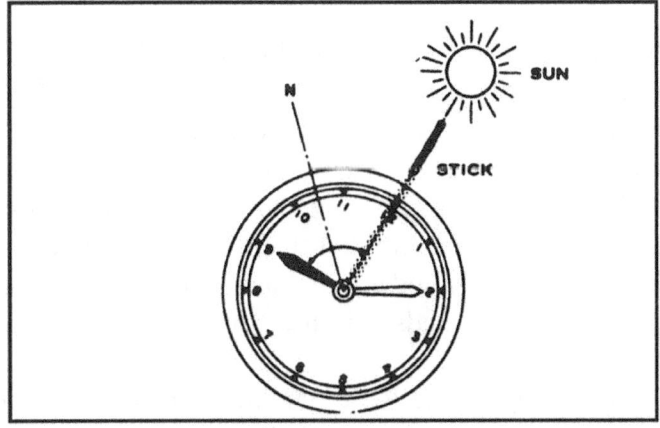

Figure 118. Finding north in the south temperate zone.

Shadow tip method for finding direction:

Drive a stake so that at least three feet of it is above the ground. Mark the tip of the shadow it casts. Wait for a while—10 minutes is long enough—and mark the spot where the tip of the shadow is then resting. A line drawn between the two marks will always point north.

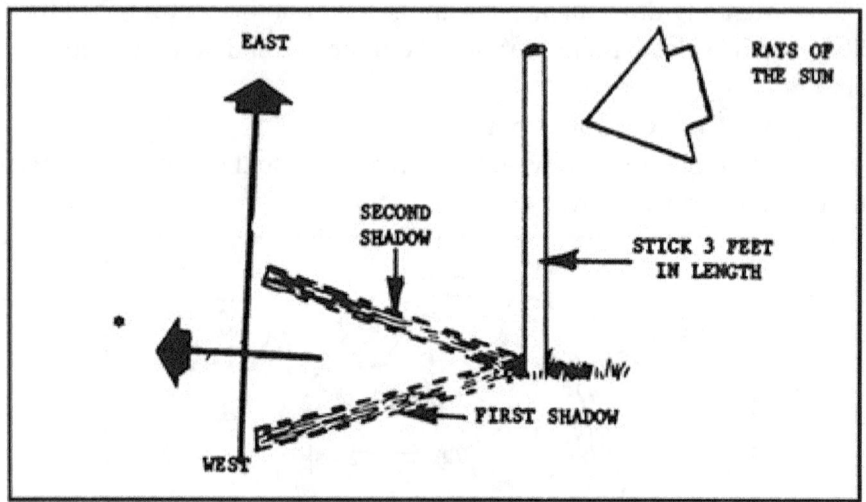

*In north temperate zone, this direction will be true north.
In south temperate zone, this direction will be south.

Figure 119. Shadow Tip Method.

Study the map. Determine the slope of the land to guide on. Notice all large waterways.

People usually live and travel on the waterways. h. Determine the direction in which you wish to go, move in one direction, but not necessarily in a straight line. Pick a linear objective, not a point objective, as it is easier to locate. Avoid obstacles—don't fight them. Take advantage of natural cover and concealment. Blundering through jungle and wooded areas leads to bruises, scratches, and quick exhaustion.

Check bearings often. Roads and trails can be used to guide on, but never travel on them. Stay alert. Natives remain on trails by preference. A few feet from the trail you are usually quite safe. Conceal yourself upon the approach of any other person until he passes or until you determine whether or not he is friendly.

The easiest traveling is often on the crests of ridges. Remember, however, that crests are more exposed than hillsides, and because of ease of travel, they are apt to be traveled more frequently than other areas.

Rivers or streams can make good roads but remember that the majority of native villages and encampments are on water. Rafts attract attention. Floating on or close to a log or drifting bush may be the simplest way to travel. Keep to the middle of the stream. If using a native boat, sink it during periods when not in use.

When close to known enemy locations, move right after sunset or just before sunrise when there is sufficient light to enable you to avoid enemy installations, mine fields, sentries, etc., but dark enough to prevent recognition by the enemy. Arrange your clothing, weapons, etc., to present a profile as similar as possible to the natives of the area.

Be quiet, noise carries far and natives are alert to any strange noise. Bury your refuse. If the enemy finds sign of your presence, it may lead to your capture.

Do not sleep near your fire or your water supply. Get far enough away to be concealed.

If lost in grass that is so tall that you cannot see over it, as a last resort cut down enough to give you some freedom of movement and, using your machete or any other tool, dig a hole to crawl into and set fire to the grass. Take every precaution not to get burned by fire or asphyxiated by smoke.

The jungle provides many hiding places. You may have to use them. Bamboo thickets are excellent. Because of the nature of bamboo, you cannot be approached without being alerted by the noise of dry bamboo.

When approaching camp, use extra precaution, for the camp is probably being watched.

At all times when hiding or remaining in one location for a period of time, be sure to plan more than one exit.

II.   SURVIVAL.

Get to a known friendly village as soon as possible. Avoid all others except as a last resort. It is difficult for a person unfamiliar with the jungle to live in it without native assistance.

Before entering any strange village, whether it is friendly or

159

not, conceal your weapons. If it is an enemy village, weapons will be taken from you. it is a friendly village, you can always go back and get them from where they are hidden.

Many of the jungle diseases are insect borne. Use insect repellent freely, if available.

Take time to repair your clothes. It helps to prevent insect bites and further tearing of clothes.

Examine your surroundings carefully. Many of your needs are there. Thorns broken from bamboo or trees can be used for needles. Strips of vines can be made into thread. If you need rope, vines will do. Your food and shelter, in fact your life, may depend on your ability to make use of things that are all a round you.

Be careful. Do not use trees and vines to pull yourself up hills as thorns, ants, scorpions, etc., will be encountered and make sores that may become infected. Use a walking stick to push aside vines and bushes.

Poisonous reptiles and large mammals of the jungle will cause few problems. Given a chance, they will avoid you.

If a survival kit is available most articles are self-explanatory. Some have multiple uses. The waterproof adhesive tape can be used for temporary repairs to clothing and mosquito nets as well as covering body wounds. Fish line can be used for snares. Three fish hooks, their shafts tied together with their hooks pointing out, can be used on the fish line to snare fish, crabs, etc. Head nets can be used as fish nets and snares. A fish hook fastened to a length of line, baited with fish or meat and left on the sea shore or in a field may be used to catch birds.

III.   WHEN REQUESTING NATIVE ASSISTANCE:

    a.   Show yourself and let the natives approach you.
    b.   Deal with recognized headman.
    c.   Do not approach groups.
    d.   Do not display weapons.
    e.   Do not risk being discovered by children.
    f.   Treat natives well. There is much you can learn from them.
    g.   Respect local customs and manners.
    h.   Learn all you can about woodcraft.
    i.   Take their advice on local hazards.
    j.   Never approach a woman.

IV.   SHELTER.

Pick a high spot when making camp. Avoid dry river beds, dead trees, and ant nests. Avoid bat caves, droppings may cause rabies.

Do not sleep on the ground if you can avoid it. Use your hammock if you have one, or make one of poncho or the multi-purpose net. If this is not possible, build a platform of bamboo, small branches, etc. It will assist in avoiding insects, reptiles, etc.

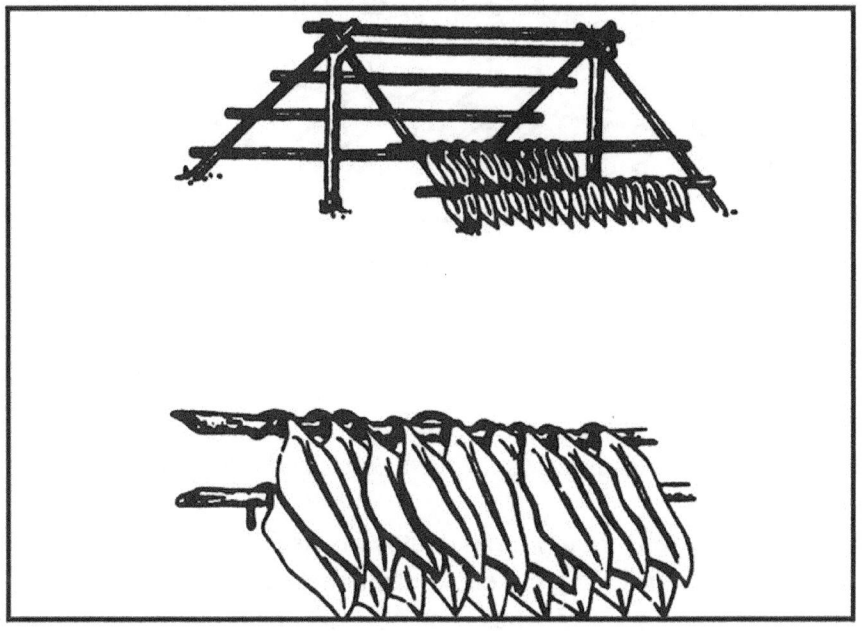

Figure 120. A-type framework.

Types of jungle shelters: Simple parachute shelter made by draping a parachute over a rope vine stretched between two trees. Thatch shelter (see figure 120) made by covering an A-type framework with a good thickness of palm or other wood leaves, pieces of bark, or mats of grass. Slant the thatch shingle fashion from the bottom upward. This type of shelter is considered ideal since it can be made completely waterproof. After finish your shelter, dig a small drainage ditch just outside its lanes and leading downhill; It will keep the floor dry.

Beds. Don't sleep on the ground; make yourself a bed of

bamboo or small branches covered with palm leaves (see figure 121). A parachute hammock may serve the purpose. You can make a crude cover from tree branches or ferns; even the bark from a dead tree is better than nothing.

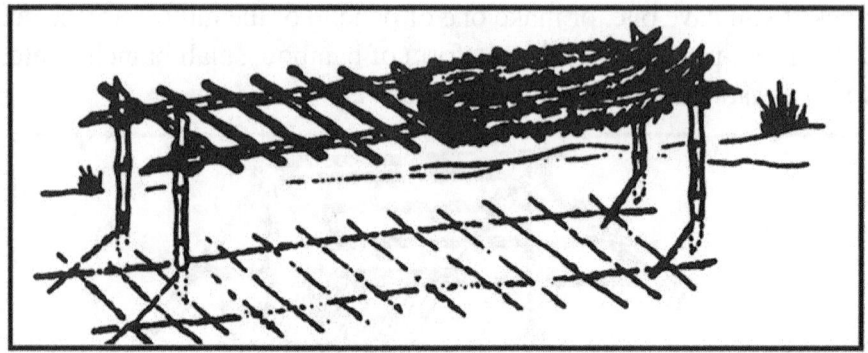

Figure 121. Bamboo Bed.

V.   WATER.

Water is more Important than food. If you have no water, do not eat. Check all drinking water for leeches and other small aquatic animals.

Indian wells. In dry areas, water can usually be found by digging a hole two or three feet deep in the bottom of dried up streams and river beds. When water has been obtained, camouflage hole.

Boiled or untreated water.

Many vines have water in them. The vine should be cut through. When a nick is cut in the vine about three feet above the original cut, a potable liquid will drip out. Do not apply vine to lips. Avoid any vine, plant, or tree with milky juice as many are poisonous. Water can be found at the base of the leaves of palms; or in sections of dead bamboo (see figure 123). A section of bamboo placed against a tree will collect water during rain. Moisture collects under leaves in the dry season. Rub these with a cloth or other absorbent material, squeeze it out into container.

At the sea shore, drinkable but brackish water can be procured by digging a hole ten feet above the high tide line.

If water is scarce, travel during coolest part of day or during night. Rest during heat of day. By doing this, the water content of the body is conserved.

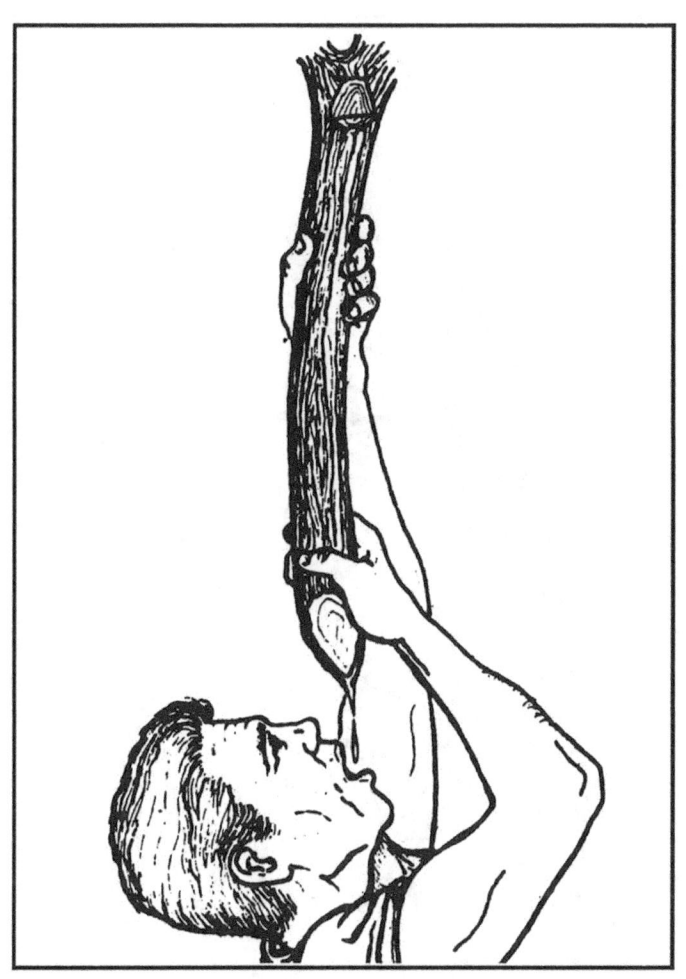

Figure 122. Extracting Water From Vines.

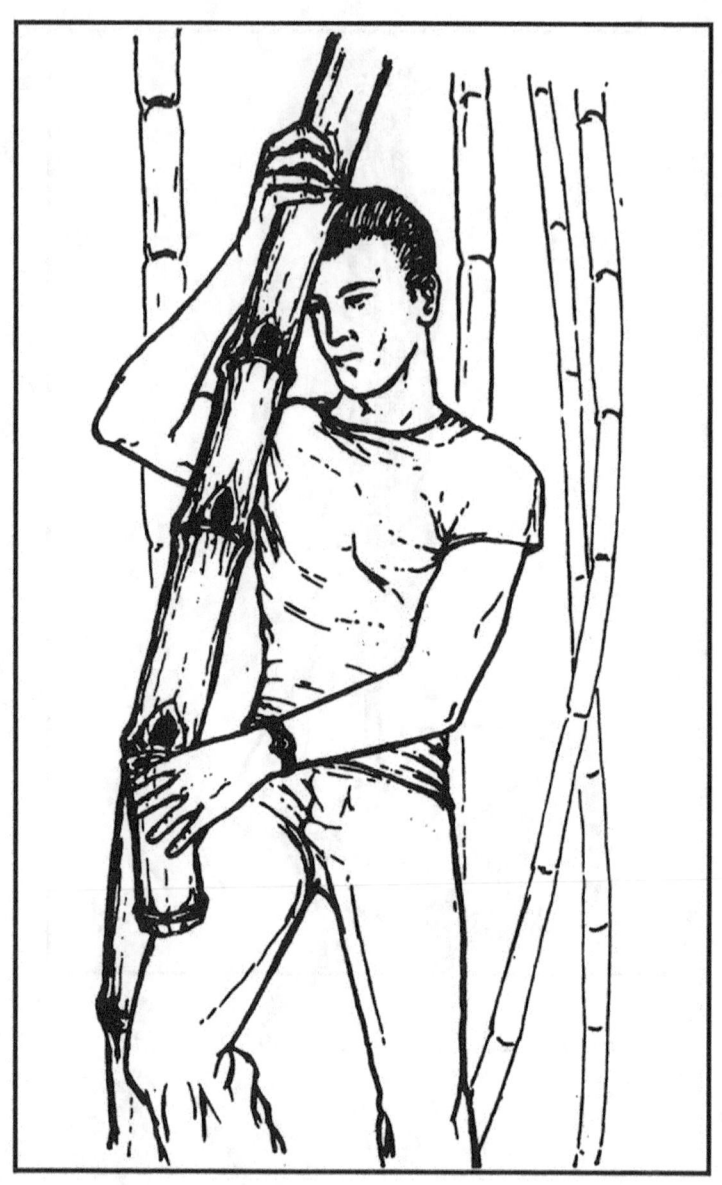

Figure 123. Bamboo joints contain water.

NO WALKING AT ALL . . .

MAX. DAILY IN SHADE TEMPERATURE (°F)	AVAILABLE WATER PER MAN, U.S. QUARTS					
	0	1	2	4	10	20
120	2	2	2	2.5	3	4.5
110	3	3	3.5	4	5	7
100	5	5.5	6	7	9.5	13.5
90	7	8	9	10.5	15	23
80	9	10	11	13	19	29
70	10	11	12	14	20.5	32
60	10	11	12	14	21	32
50	10	11	12	14.5	21	32

MAX. DAILY IN SHADE TEMPERATURE (°F)	AVAILABLE WATER PER MAN, U.S. QUARTS					
	0	1	2	4	10	20
120	1	2	2	2.5	3	
110	2	2	2.5	3	3.5	
100	3	3.5	3.5	4.5	5.5	
90	5	5.5	5.5	6.5	8	
80	7	7.5	8	9.5	11.5	
70	7.5	8	9	10.5	13.5	
60	8	8.5	9	11	14	
50	8	8.5	9	11	14	

WALKING AT NIGHT UNTIL EXHAUSTED AND RESTING THEREAFTER

NOTE: Columns 2–7 show survival time in days.

VI.   FOOD.

There is food in the jungle if you know where to find it. Plan one good meal each day but nibble on any food that you may have or can find. Eat strange in small quantities and wait for a reaction. Avoid all mushrooms. There is little nutritional value in them and much danger.

In villages, eat only food that is hot, if possible. If for fear of offending your host you have to eat native food that is not hot, take a yellow pill to avoid dysentery. All vegetable or fruit procured in a village or handled by natives should be peeled.

Possession of a knife is vital for successful foraging. If you do not have one, a serviceable blade can be made from split bamboo. Split dry bamboo with a stone, break out a piece, sharpen on a stone, fire harden and resharpen. The result will be a crude but effective tool or weapon.

Animal food. Grasshoppers, ant eggs, hairless caterpillars, larvae and termites, are good when cooked. Remove heads, skin, and intestines of snakes, rats, mice, frogs, lizards, before cooking. Bats can be caught in caves by flailing the air through which they are flying with a multi-branched stick. Inasmuch as bats are carriers of hydrophobia, do not get bitten.

Traps and snares. Indiscriminate placings of traps is a waste of time. Small game such as rabbits, mice, etc., travel on paths through the vegetation. Set traps in or over these trails. A serpentine fence will guide certain birds, like pheasants and some larger animals, to your traps. Cut or collect brush for the fence and build it two feet high or more. Place traps in depth of curve.

Figure 124. A simple deadfall using a figure 4 trigger.

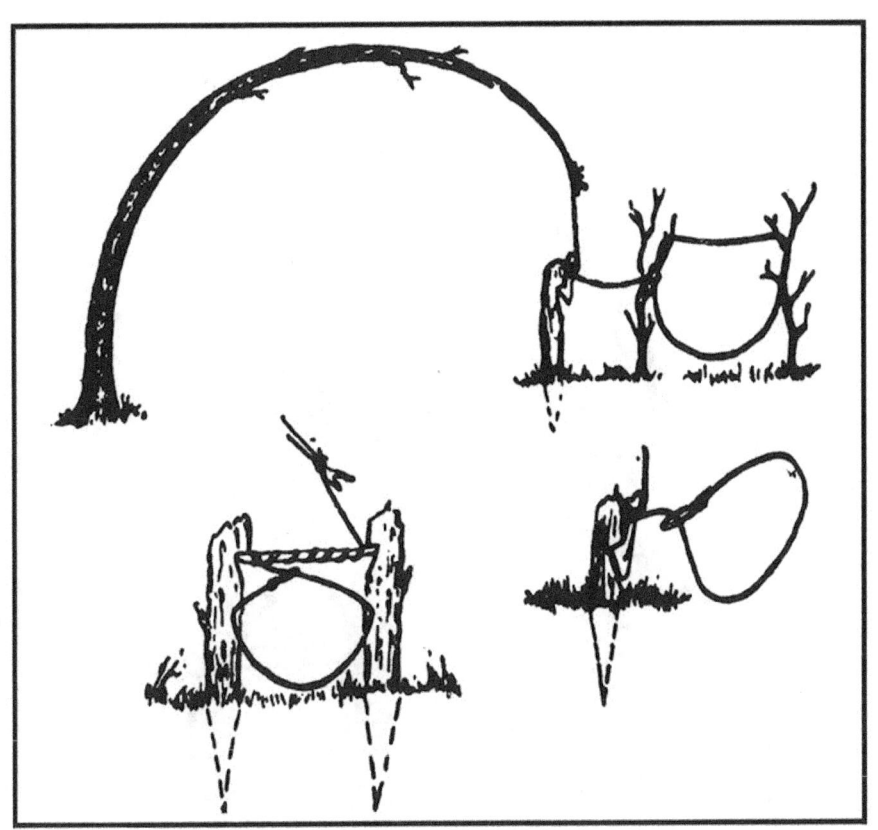

Figure 125. Fixed snares.

Figure 126. Hanging snares.

Fish. There is no rule to determine edible fish. Avoid all strange or oddly shaped fish. Only those mussels, clams, oysters, etc., that are found underwater at low tide are safe. Salt water fish and shell fish can be eaten safely raw. Do not eat the eggs or intestines of any fish. Salt water snails come in all sizes and shapes. All are good to eat. Avoid cone snails and terebra. Some have poisonous stings that can be fatal. Never eat fresh water fish without cooking or when the flesh is soft or the eye sunken for they are undoubtedly diseased.

Fish are attracted to light. If the area is safe, use torches at night to attract the fish. A head net made in a circular form by threading with bamboo or strung on a crotched stick will make a dip net. Fish

168

in ponds or at the edge of the beach can be driven into the shallows by flailing the water with hands or brush. Clean fish immediately when caught. If you are in a group, work together to drive the fish and to net them. Help each other.

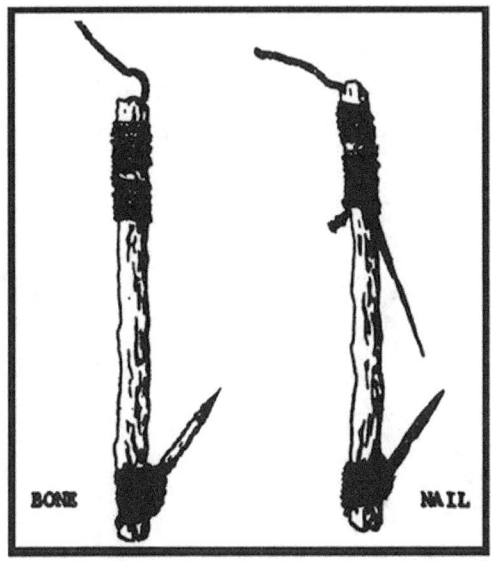

Figure 127. Improvised hooks and lines.

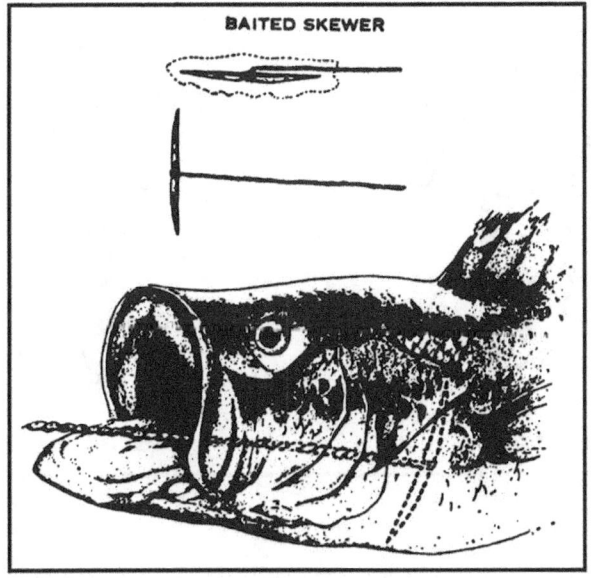

Figure 128. Skewer hook.

Do not try to preserve meat or fish for any length of time. In the tropics flesh of any kind spoils rapidly unless dried or smoked.

Skinning and cleaning. As soon as you catch a fish cut out the gills and large blood vessels that are next to the backbone. Scale it. Gut the fish by cutting open its stomach and scraping it clean. Cut off the head unless you want to cook the fish on a spit. Fish like catfish and sturgeon have no scales. Skin them. Small fish under four inches require no gutting, but should be scaled or skinned.

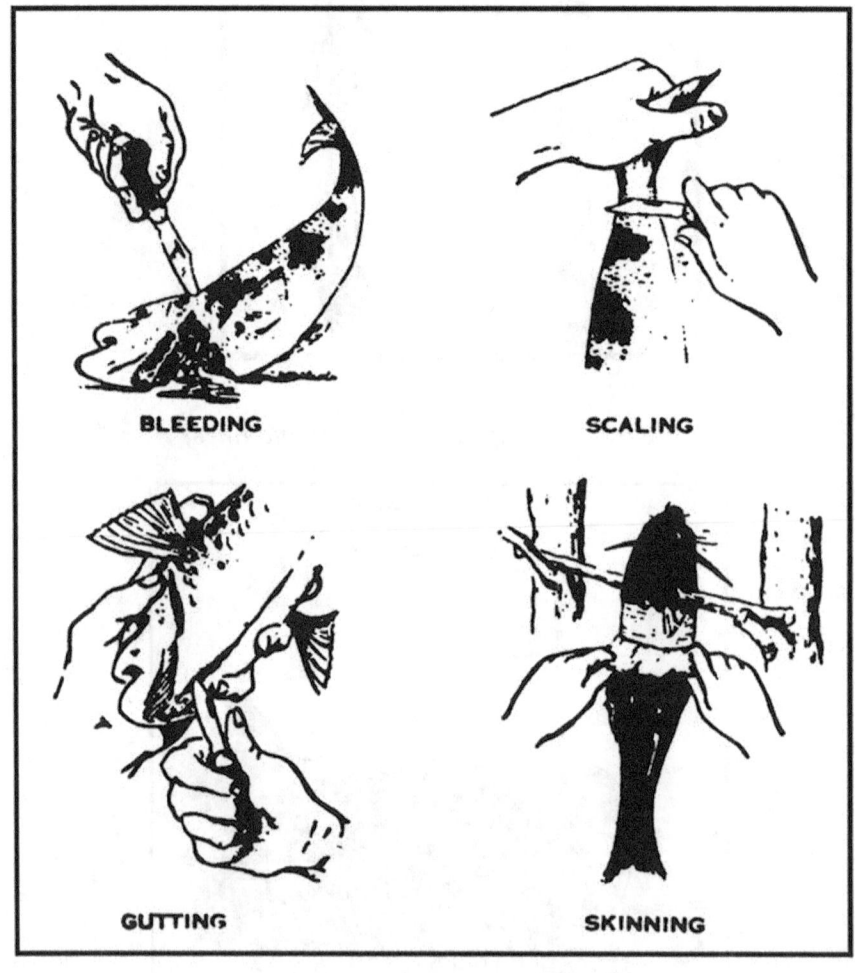

Figure 129.

VII.   FIRE.

Keep your fire small. In the rainy season or in damp jungles, dry fuel may be difficult to obtain. Carry dry tinder with you to assist in starting your fire. By cutting away the wet outer cover of a sound log, dry fuel can be obtained. Shave dry wood or dead bamboo into thin slivers and stack in tent formation over tinder. Pile heavier fuel around fire and add slowly until fire is well started. If fuel is damp, stack it close to fire to dry out.

If the jungle floor is flooded or may become so, build your fire on a hearth of stones or wet wood. If necessary, build a shelter over the fire to protect it from the rain. If the weather gets cold and you need fire for survival, build a screen on the opposite side of the fire from you to reflect the heat toward you. A screen of leaves or branches three or four feet square tied together with fish line or vines will do the job. Tilt the screen with the top toward you. Fiber soaked in insect repellent makes good tinder.

VIII.   COOKING.

If larger game has been killed, the stomach or skin can be made into a cooking vessel after being cleaned. Fasten three strings into holes made in the top of the wall of the open stomach or skin pouch and tie to the apex of a tripod made of sticks. Fill with water, which can be brought to a boil by putting in fire-heated stones. If sticks are not available and if the ground is not too wet or stony, the skin or stomach pouch can be used as a liner for a bole in the ground. Then fill with water and place fire-heated stones into it.

Meat and fish can be stuck onto a sharpened green stick and roasted over a fire.

Small animals and birds can be roasted easily. Draw and skin them and wrap in leaves, clay, or mud. Bury in a pit, the bottom of which is lined with heated stones. Fill pit with dirt. In the morning when the pit is opened, you will find the meat well cooked and hot. Larger game can be prepared the same way by cutting into small pieces.

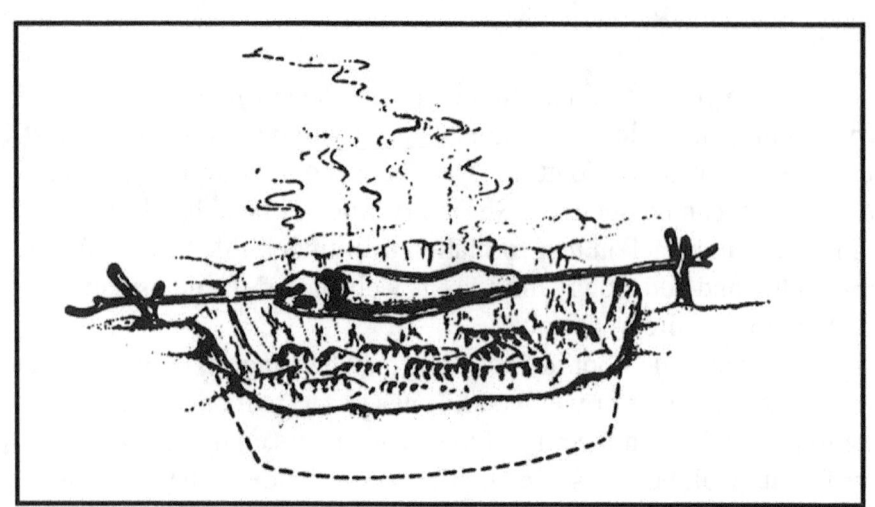

Figure 130. Pit fire.

Figure 131. Simple crane.

IX.   HEALTH.

Care of your person is extremely important. If you have a survival kit, directions for the use of drugs are printed on the containers.

Treat every wound or sore as soon as possible. To stop bleeding in the absence of bandages, apply freshly made spider webs. This will assist in the coagulation of the blood.

In the absence of toilet paper, use leaves and grasses. Be careful to examine the leaves and grasses for insects. Use no leaves that have any fussy or hairy surfaces or are taken from a tree or plant with milky sap, or grass that has a serrated edge. Do not use material that is laying on the ground.

Leeches and ticks can be partially avoided by tying cuffs of your jacket at the wrist and the bottoms of trouser legs outside the boots and applying insect repellent to all openings. Check your clothes and body frequently. Remove leeches and ticks carefully. If pulled off quickly, they may leave their heads in the bite. Infection will result. Wet salt, fire, or lime juice will cause them to withdraw their heads and fall off. Don't hurry the process.

In case of heat stroke, heat exhaustion or heat cramps, lower the body temperature by drenching with water or covering the body with wet clothing. Dissolve two salt pills in the equivalent of a cup of water and drink. Rest until all symptoms have passed.

In cases of diarrhea when no drugs are available, a tea made from boiled guava leaves or charcoal eaten with hot water will be beneficial.

Boils can be brought to a head by applying hot pads.

Avoid sunburn. Even a short time in the jungle will reduce your resistance to the sun. Serious infection can result from over-exposure. Keep covered. Do not risk a painful, dangerous burn.

MOST IMPORTANT OF ALL, KEEP YOUR HEAD, TRY NOT TO GET TOO TIRED, REST FREQUENTLY, BE CAREFUL, AND DO NOT GIVE UP.

CHAPTER 9
MISCELLANEOUS

CONVERSION TABLE-WEIGHTS AND MEASURES.

MULTIPLY	BY	TO OBTAIN
Acres	.405	Hectares
Caliber (inches)	25.4	Millimeter
Centimeters	.3937	Inches
Degrees	17.8	Mils
Fathoms	6	Feet
Feet	.1667	Fathoms
Gallons (US)	3.785	Liters
Grains	.00228	Ounces
Grams	.03527	Ounces
Hectares	2.471	Acres
Inches	2.54	Centimeters
Kilograms	2.2	Pounds
Kilometers	.6214	Miles
Knots	1.152	Miles per hour
Liters	.2642	Gallons (US)
Meters	1.094	Yards
Miles	1.609	Kilometers
Miles per hour	.8684	Knots
Millimeter	.0394	Inches
Mils	.056	Degrees
Ounces	437.5	Grains

TABLE - AERIAL PHOTOS.

Determining Scale:
> Flying Height Method:

$$\text{Scale} \quad \frac{F \text{ (Focal length of camera in inches)}}{H \text{ (Altitude above ground in inches)}}$$

> Map Distance Method:

$$\text{Scale} \quad \frac{FD \text{ (Photo distance in inches)}}{GD \text{ (Map distance in inches)}}$$

Point Designation Grid System

1. Turn photo so that written data is in normal reading position.

2. Draw lines across photo joining opposite fiducial (collimating) marks.

3. Space grid lines, starting with center lines, a distance equal to 4 cm or 1.576 inches apart.

4. Number each center line 50 and give numerical values to the other lines, increasing right and up.

5. Read coordinates as any other.

TABLE LONG RANGE PHOTOGRAPHY.
(35-mm Camera & Binoculars)

Procedure.
> Camera:

F Stop -	M 13 6 × 30 binoculars F 10
	M 17 7 × 50 binoculars F 8
Speed -	As required by film ASA
Range -	Infinity

Binoculars:
>
> Set left eyepiece at zero.
>
> Sight through right eyepiece and adjust to focus.

Set binoculars to camera:

> Place left monocular (with reticle) flush with camera lens.
>
> Take picture without moving either binoculars or camera.

TABLE - MAP-DISTANCE CONVERSION

Map distance	Ground distance	1/25,000	1/50,000	1/75,000	1/100,000	1/200,000	1/250,000	1/500,000	1/1,000,000
One inch	Inches	25,000	50,000	75,000	100,000	200,000	250,000	500,000	1,000,000
	Feet	2,083	4,167	6,250	8,333	16,667	20,833	41,667	83,333
	Yards	694	1,389	2,083	2,778	5,555	6,944	13,888	27,776
	Meters	635	1,270	1,905	2,540	5,080	6,350	12,700	25,400
	Miles	0.4	0.8	1.2	1.6	3.2	4	8	16
	Kilometers	.64	1.3	1.91	2.54	5.08	6.35	127	25.4
One centimeter	Inches	9,843	19,685	29,528	39,370	78,740	98,425	196,850	393,700
	Feet	820	1,640	2,460	3,281	6,562	8,202	16,404	32,808
	Yards	273	547	820	1,094	2,187	2,734	5,468	10,936
	Meters	250	500	750	1,000	2,000	2,500	5,000	10,000
	Miles	0.16	0.3	0.5	0.6	1.2	1.5	3	6
	Kilometer	.25	.50	.75	1.00	2.00	2.50	5.00	10.00

Representative fraction (RF)

TABLE - USEFUL KNOTS.

Name	Illustration	Use
Square		Join two ropes of same size. (Will not slip, but will draw tight under strain.) To and block lashing.
Double sheet bend		Join wet ropes, of unequal size, or rope to an eye. (Will not slip or draw tight under strain.)
Bowline		Form a loop. (Will not slip under strain and is easily untied.)
Timber hitch		Lifting or dragging heavy timbers. (Is more easily controlled if supplemented by half hitches.)
Clove hitch		Fasten rope to pipe, timber, or post. (It is used to start and finish all lashings and may be tied at any point in rope.)
Sheep shank		Shorten rope or take load off weak spot in rope.
Anchor knot		To fasten cable or rope to anchor.

TABLE - MISCELLANEOUS INFORMATION.

PRINCIPLES OF WAR	REPORTING INFORMATION
M ass	S ize
O bjective	A ctivity
S implicity	L ocation
S urprise	U nit
C ommand unity	T ime
O ffensive	E quipment
M aneuver	
E conomy of forces	
S ecurity	

TERRAIN ANALYSIS	PRISONERS OF WAR
C ritical features	S earch
O bservation	S eparate
C over and concealment	S ilence
O bstacles	S peed
A venues of approach & withdrawal	S afeguarding

INTELLIGENCE EVALUATION LEGEND.

Source	Information
A-Completely reliable	1-Confirmed by other source
B-Usually reliable	2-Probable true
C-Fairly reliable	3-Possibly true
D-Not usually reliable	4-Doubtfully true
E-Unreliable	5-Improbably
F-Reliability unknown	6-Truth cannot be judged

This legend should be applied to intelligence originating in the field and the evaluation sent forward with the information.

GUERRILLA TRAINING

I. GUERRILLA TRAINING AIMS: Survive, Obey, Fight.

II. TRAINING PLAN:

Steps in planning:

1. Analysis of the mission.

2. Systems for training:
   a. Decentralized.
   b. Centralized.
   c. Combination of Systems.

3. Estimate of training situation:
   a. Training to be conducted.
   b. Personnel:
      1. Available for cadre.
      2. To be trained.

4. Time.

5. Training facilities.

6. Training aids.

7. Equipment.

Decisions.

The Plan.

Principles of scheduling:

1. Facilities preparation of instruction.
2. Facilities learning.
3. Use training time effectively.
4. Accommodate the troops.

III.   LEGAL STATUS OF GUERRILLAS:

Be commanded by a person responsible for his subordinates.
Have a fixed distinctive insignia recognizable at a distance.
Carry arms openly.
Conduct operations in accordance with the laws and customs of
war.

IV.   FOR SUCCESSFUL EMPLOYMENT OF GUERRILLA
WARFARE:

The spirit of resistance must be present in a segment of the
population.
The guerrillas must have the support of the civilian populace.
The guerrilla movement must have a sponsor.

V.   RECORDS OF GUERRILLAS:

Personnel roster; name, rank, date joined, date discharged.
Oath of enlistment.
Theatre records and reports.
Casualty reports.
Payrolls.
Recording and settling claims.
Receipt forms
Demobilization:
1.   Assembly of guerrilla forces.
2.   Collection of arms and equipment.
3.   Completion of administrative records.
4.   Settlement of pay, allowances, and benefits.
5.   Settlement of claims.
6.   Awarding of decorations.
7.   Care of sick and wounded.
8.   Discharge.
9.   Provisions of rehabilitation and employment of
discharged guerrillas.

VI. GUIDE TO ASSESSMENT OF THE AREA:

Initial Assessment.

1. Location.
2. Team morale and condition.
3. Status of guerrillas (local).
4. Security (local): area, attitude of local civilians, escape plan and alternate areas, enemy situation, civilian support available.

Principal Assessment (a continuous estimate of the situation).

1. Information of the enemy to include: Disposition; composition, identification, and strength; organization, armament, and equipment; degree of training, morale, and combat effectiveness; operations (recent and current activities of the unit, counterguerrilla activities, and capabilities, current security systems within the unit); unit zones of responsibility; daily routine of the units; logistical support to include: installations and facilities, supply routes, method of troop movement; past and current reprisal actions.

2. Information of security troops and police units: Dependability and reliability to the existing regime and/or the occupying power; disposition; composition, identification, and strength; organization, armanent, and equipment; degree of training, morale, and efficiency; influence on an relations with the local population.

3. Information of resistance organization: Size, equipment, organization, status of training, intelligence and logistics systems. Auxiliary, organization, status of training, general disporitions.

4. Information of the civil government: Controls and restrictions (documentation, rationing, travel and movement restrictions, blackouts and curfews); current value of money, wage scales; the extent and effect of the black market; political restrictions; religion restrictions; the control and operation of industry, utilities, agriculture, and transportation.

5. Information of potential targets: Railroads; telecommunications ; POL; electric power; military headquarters and instal-

lations; radar and electronic devices; highways; inland waterways and canals; sea ports; natural and synthetic gas lines; industrial plants.

6.   Information of the terrain: Location of area suitable for guerrilla bases, units and other Installations; potential landing zones, drop zones, reception sites; routes suitable for guerrillas and enemy; barriers to movement; the seasonal effect of the weather on terrain and visibility.

7.   Information of the weather: Precipitation, cloud cover, temperature and visibility; wind speed and direction; light data (BMNT, EENT, sunrise, sunset, moonrise and moonset).

Special Forces Prayer

Almighty GOD, Who art the Author of liberty and the Champion of the oppressed, hear our prayer.

We, the men of Special Forces, acknowledge our dependence upon Thee in the preservation of human freedom.

Go with us as we seek to defend the defenseless and to free the enslaved.

May we ever remember that our nation, whose motto is "In God We Trust," expects that we shall acquit ourselves with honor, that we may never bring shame upon our faith, our families, or our fellow men.

Grant us wisdom from Thy mind, courage from Thine heart, strength from Thine arm, and protection by Thine hand. It is for Thee that we do battle, and to Thee belongs the victor's crown.

For Thine is the kingdom, and the power and glory, forever,

AMEN.

www.ingramcontent.com/pod-product-compliance
Lightning Source LLC
Chambersburg PA
CBHW071716170526
45165CB00005B/2030